D0471231

The NORTH AMERICAN DESERTS

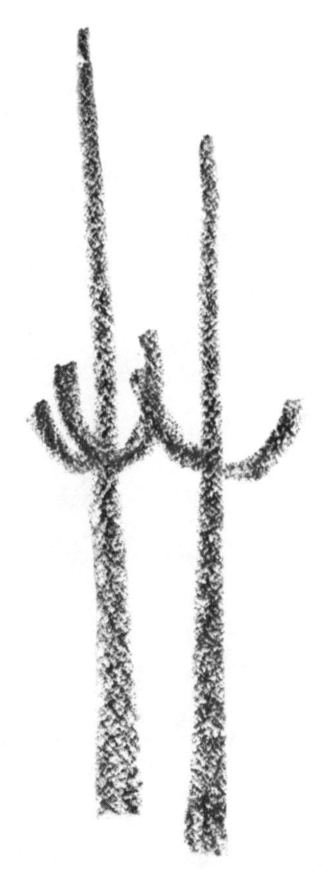

The NORTH AMERICAN

DESERTS

by EDMUND C. JAEGER

With a chapter by PEVERIL MEIGS *and illustrations by* JOHN D. BRIGGS, LLOYD MASON SMITH, MORRIS VAN DAME, *and the author*

STANFORD UNIVERSITY PRESS • Stanford, California

ISBN 0-8047-0498-8 Original edition 1957
Last figure below indicates year of this printing:
83 82 81

Title page: Forest of *Idria columnaris,* in the Vizcaíno Desert
of Baja California, Mexico (photograph by
Dr. Homer Ashmann)

PREFACE

Each year thousands of persons pass by train, bus or private automobile through our deserts. To the greater number, these arid lands are but places desolate and inhospitable, great expanses of monotony against waterless horizons. They think of deserts as singularly devoid of animal life, "except, of course, lizards, snakes, scorpions, and spiders," and with only "sage and greasewood bushes and cacti" here and there to break the monotony of gray sand and rock. Actually thousands of wonders lie about them, often within stone's throw, never-to-be-forgotten sources of surprise, admiration, and pleasure that can be theirs to enjoy if only they expend a little effort to recognize and interpret what they see.

Besides these desert visitors are those who find themselves seeking the desert by road and trail for combined recreation and serious study of the natural history. Others there are who have become permanent desert residents and who know and love the deserts mostly as a tranquil land of sunshine by day and near-cloudless skies lit by brilliant stars at night. Some of these, I feel certain, need to be made more aware of the desert's rich heritage of natural beauty and out-of-door phenomena.

The happiest travelers, the most fortunate desert dwellers are those who are constantly curious about their ever changing surroundings and who are always eager to discover something more definite about the strange things they see. It is to these ever alert questioners that this book is addressed.

The desert abounds in so many unusual things to be observed that it becomes almost imperative that those who are largely uninitiated in desert lore should first learn a few of the desert's fundamental characteristics; then later they will discover its more subtle scenic aspects and its less easily seen inhabitants.

Here for the first time is presented a comprehensive yet simple picture of all of our North American deserts, deserts five in number which together extend from central Mexico almost to the border of Canada. Maps show their broad extent. Numerous illustrations (photographs and line drawings) reveal their natural beauty and their typical animal and plant inhabitants.

Many phases of desert phenomena are introduced so that not only may the picture be well rounded but also that there may be something to appeal to persons of different interests. If it should occur to anyone that botanical features have received undue attention, let him remember that the trees, shrubs, and flowering annual plants are the most obvious and perhaps best indicators of the nature of the environment. To each bird, lizard, or insect he notices there are in comparison thousands of plants which pass before his eyes and attract his attention. No one can fully appreciate the desert around him if he does not somewhat intimately know the plants that not only form the ground cover but also furnish food, shelter, and protection from enemies for the desert's numerous animal denizens.

Although the presentation is scientific much of the phraseology of science has been purposely avoided; but scientific names have not been omitted for they form an accurate means of identification for the serious reader as well as a helpful and easy way, through association, of familiarizing the young student with them. Selected lists of references are given for readers desiring more detailed information about the numerous items mentioned.

It is suggested that both desert dwellers and wayfarers keep this guide to desert knowledge close at hand. During odd moments they should thumb through the pages to become thoroughly familiar with the illustrations so that they will have some idea of what to look for as they travel. The numbers in parentheses after plant and animal names refer to illustrations in the sections following Chapter 14.

Alongside the names and descriptions accompanying the line drawings of animals and plants are marginal blank spaces. These are good places to indicate by an appropriate mark such as a check (√) or cross (×) that specimens have been found and recognized. It will add enormously to one's interest to see how many of the desert's plant and animal names can thus be marked: indeed it makes an enticing game to play while traveling along roads or footpaths or when camping in wilderness areas.

The author is indebted to many persons who aided by criticisms, supplying information and illustrations, or giving travel directions which aided in his explorations. Among these are Lloyd Mason Smith, former director of the Palm Springs Desert Museum, who gave invaluable help in selecting material for many of the chapters and, with Charles S. Papp and Morris Van Dame, made drawings of the mammals, birds, and reptiles; Homer Aschmann and Herman Spieth of the University of California at Riverside; Carl Hubbs of the Scripps Institution of Oceanography; Charles Shaw, L. M. Klauber, and the late C. B. Perkins of the San Diego Zoological Society; Ira Wiggins and Roxana Ferris of Stanford University; George Lindsay of the San Diego Museum of Natural History; Samuel King, former Superintendent of the Joshua Tree National Monument; Harold S. Colton, director, and Malcolm F. Farmer, assistant director, of the Museum of Northern Arizona; Lyman Benson of Pomona College; John Briggs of the University of Illinois, who made the insect drawings; Cornelius H. Muller of the University of California at Santa Barbara; William H. Woodin, director of the Arizona-Sonora Museum; Samuel Dicken of the University of Oregon; Angus M. Woodbury, director of Ecological Research at the University of Utah; Paul D. Hurd of the University of California at Berkeley; and Stillman S. Berry of Redlands. Clyde Butler made the regional maps. Credit lines have been added beneath the photographs so kindly provided by professional friends and institutions. Photographs without credit lines are by the author.

Nor would I fail to mention the many citizens of Mexico who always so eagerly and courteously helped me to appreciate the fascinating desert land below the International Border. Among these were fishermen, farmers, *vaqueros,* soldiers, policemen, and shopkeepers.

EDMUND C. JAEGER

Riverside, California
May 15, 1957

CONTENTS

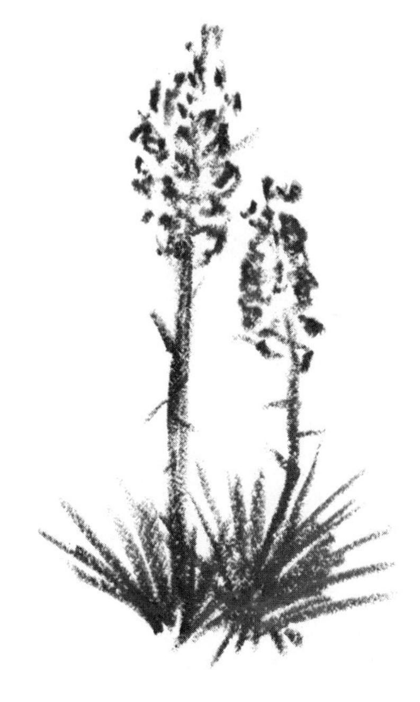

There is something infectious about the magic of the Southwest. Some are immune to it, but there are others who have no resistance to the subtle virus and who must spend the rest of their lives dreaming of the incredible sweep of the desert, of great golden mesas with purple shadows, and tremendous stars appearing at dusk from a turquoise sky. Once infected there is nothing one can do but strive to return again and again.

From PREHISTORIC INDIANS OF THE SOUTHWEST,

by H. M. Wormington

1

WHAT IS A DESERT?

Nearly one-fifth of the surface of the earth is made up of deserts, supporting less than four per cent of the world's population. Although individual parts of these arid regions are quite different in physical appearance, they possess in common several characteristics, such as low rainfall, high average temperatures during the day, and almost constant winds, with consequent increased rate of evaporation.

The most important factor in the creation of a desert is a low annual rainfall. Most geographers have arbitrarily agreed that if a region receives less than ten inches of unevenly distributed rain throughout the year, then it may be termed a desert. The area in question must have in addition a relatively high mean yearly temperature. It is obvious that it would be impossible for a true desert to exist in very cold climates where most of the moisture that falls is retained by freezing and never actually lost. Cold barren regions such as are found in the Arctic and Antarctic can be called wastelands, but certainly not deserts in the true sense.

Because of the high average temperature there is rather rapid evaporation of the little rain that does fall, so that during any year only a small amount of moisture is available for the animals and plants. Naturally the time of the year in which the normal rains occur is important, for if these are mostly in summer, when water loss through evaporation is great, that particular area will be much more arid than one which receives equal amounts of rain in cooler weather.

Much of the summer rain of deserts may be of the cloudburst type, with perhaps several inches falling in a few hours. Most of this moisture is lost to the plants because of the rapid surface run-off. Such rains may cause destructive sheet floods, which are responsible for the burying of young plants under sand and the undercutting of root systems of both immature and mature ones.

Desert areas usually have winds almost constantly sweeping

across them, winds that dry out both soil and vegetation. The prevailing winds generally blow *into* the desert from its fringes. The heavy cold air moving into the area replaces the light heated air rising from the desert floor. If the incoming wind is channeled through mountain passes it may be a very steady one and often of considerable force. Such a wind both dries out the country and materially aids the rains in the shifting of soil, causing erosion. The amount of material annually transported by desert winds can be very great.

Almost all deserts are considerably lower in elevation than their surrounding mountains. The streams emptying into them are few. Because most deserts are basins without outlets, the water that collects in their lowest parts soon becomes quite alkaline, and finally, through rapid evaporation, disappears, leaving behind dry lake beds or "clay pans," some of which become heavily encrusted with salts.

Since the plant cover of deserts is necessarily sparse and the amount of solar radiation from the surface soil is great, little of the diurnal heat is retained after the sun goes down. This accounts for the cool, if not cold, nights that often follow very warm days. Because extremes in temperature between day and night are often remarkably high, as much as 70° or 80° F. within a few hours, the animals and plants must be specially adapted to withstand such rapid fluctuations.

Among the causes of deserts are mountains which act as barriers cutting off the moisture-laden clouds that might otherwise sweep across and deposit rain upon them. As moisture-filled clouds blow in and rise they become chilled, and now, no longer able to retain their load of moisture, drop it as rain on the windward side of the mountains away from the desert.

In a few cases, such as in the Atacama Desert of coastal Peru and the Kalahari and Karoo deserts of South Africa, a cold ocean current acts much as mountains do in robbing the rain clouds of moisture.

Each of the continental land masses has its desert. By far the largest in extent is the great Eurasian Palearctic Desert, which includes the Sahara and the deserts of Asia Minor, India, Tibet, China, and Mongolia. Most of the interior of Australia is desert, local names of its parts being the Arunta, Gibson, and Great Victoria deserts. Another major desert area includes much of the southwestern portion of the United States and north-central and northwestern Mexico. For convenience in description, this American Desert region may be divided into five subareas: The CHIHUAHUAN, SONORAN, NAVAHOAN, MOHAVEAN, and GREAT BASIN deserts. Of these the Sonoran is sufficiently diversified to warrant a further cleavage into six subdivisions: the SONORA PROPER, ARIZONA UPLAND, YUMAN, COLORADO, VIZCAINO-MAGDALENA, and GULF COAST deserts. In this book each of the North American desert units is treated in a separate chapter.

Plants which have become specially adapted to desert conditions are called xerophytes, a word derived from Greek words meaning "dry plants." In order to be able to withstand great heat and severe drought, often over long periods lasting months and occasionally even several years, xerophytic plants have adjusted themselves to their environment in several remarkable ways. They may survive the hot summers by then remaining in the seed stage. They sprout forth only during the rainy season, then grow rapidly, flower, and go back to the seed stage with the advent of the dry season. Most of the small desert annuals are of this type. Their seeds may, if necessary, lie dormant, while buried in the soil, for many years, waiting until a propitious time to begin growth.

Desert plants can also survive by passively evading the dry months. This they do not only by storing and gradually utilizing the moisture received during the year but also by maintaining throughout their lifetime a dwarf form. Such plants are usually widely spaced. Because of their small size they do not need much water to begin with and seem to get along quite well, even to flourish, with unbelievably small amounts of it. Some of the dwarf fragile-stemmed wild buckwheats (*Eriogonum*) are familiar examples.

During dry spells plants may go into a state of dormancy, suspending all normal activity. Such a state is called "drought endurance" by the plant physiologists. Only the advent of rain will arouse such plants to activity. Once the rains come they often leaf out within a period of days. Later the leaves may be almost as quickly shed and there is a sudden return to dormancy. Ocotillos, elephant trees, and jatrophas are desert plants subject to such fluctuations of activity.

Another adaptation is found in those plants which actively resist aridity by storing water within their leaves and stems or roots. They are able to continue growing even through the hottest months when all other desert vegetation suspends almost all growth activity. Many such plants attain to larger than usual bulk and stature and at the same time develop long lateral roots for quickly utilizing surface moisture and long tap roots to reach deep sources of soil water. The various kinds of cacti are examples of such water storers and conservers. Especially typical are the sahuaro and the barrel cacti. During the rainy season the columnar fluted stems become almost "bloated" with water. As the season gets progressively drier and the stored water is gradually used up, the accordion-like stems shrink and become very lean.

When leaves are present on desert plants they are usually greatly modified for water conservation. It is through the leaf that most of the plant's moisture is lost by evaporation. During the warmest part of the day xerophytes often turn or twist their leaves so that only the thin edges are exposed to the sun's direct rays. Leaves that cannot be turned or twisted may curl or roll up during the hotter part of the day and then uncurl in the cooler hours of later afternoon and early morning.

Many desert plants have exceedingly hairy stems and leaves which quickly catch and retain moisture of the surrounding air; the sand verbena is an example. The same hairs may shield the stem and leaf surfaces from direct sun exposure. Leaves of some desert plants have a thick, leathery epidermis to protect them against too rapid water loss; still others, such as the creosote bush and varnish-leaf acacia, have a shiny waxy coating which reflects heat.

Even the breathing pores (stomata) of the leaves of xerophytes exhibit special adaptations. Although many in number, they are typically quite small and either sunken in hair-lined cavities for added protection or shielded by waxy secretions. Further, they may be equipped with valves which close during the day. Normally these vital breathing pores are located on the underside of the leaf away from the sun.

A number of desert plants, such as the smoke tree, crucifixion thorn, and the cacti, have given up leaves almost entirely and have modified their stems to take over leaf functions. In these the outermost coating of the stem, the cuticle, has become toughened and thickened, not only as a means of conserving water but also as a shield against injury and the etching action of wind-borne sand.

A surprisingly large group of xerophytic plants have developed sharp spines or stiffish hairs of one sort or other. Most cacti have them, as do also the acacias, mesquites, condalias, ocotillos, yuccas, and smoke trees. Just why these particular plants have developed a thorny armor is somewhat conjectural, but thorns

Cardon Forest of giant cardon, mid-region of the Vizcaíno Desert, Baja California. The smaller cacti are species of opuntia and machaerocereus. Volcanic hills in background.

Oregon State Highway Commis

Left: U.S. Highway 395 from Canada to Mexico, called the Three Flags Highway, takes a straightaway through Oregon's high plateau rangelands in the central part of the state near Wagontire. The Old West is suggested through this country by sagebrush, vast cattle ranches, and bands of range horses.

Lower left: Lush shrub-and-tree desert near Sonoyta, Sonora, with organ pipe cactus (*Cereus thurberi*) and brittle bush (*Encelia farinosa*) the most conspicuous plants. Palo verde trees (*Cercidium floridum*) in background.

Below: Beautiful mountains of banded limestones of the southern Nevada deserts. In the flatlands are vast forests of tree yuccas (*Yucca brevifolia jaegeriana*). The summers are characterized by high diurnal temperatures and steady hot winds. Snow sometimes blankets the land during the season of winter storms.

L. Burr Belden

Avery Edwin Field

Bullion Mountains, Mohave Desert, with sand blown upon their flanks. Flats covered with creosote bush (*Larrea divaricata*) and the low, rounded, gray-green burro bush. The clouds are typical of those which gather during the hot days of summer.

certainly do, in most cases, act as a protection against being eaten. In addition to spines, some desert plants have developed pungent-odored, bad-tasting, or poisonous substances which deter hungry animals from eating them.

In adapting themselves to severe conditions it is ordinarily the stems and leaves of xerophytes that have changed most of all. The roots have suffered little structural change but they may have altered their manner of growth and distribution. Most of the desert plants have root systems consisting of numerous laterals well spread out and growing close to the surface in order to take quick advantage of the shallow-penetrating rains. A few xerophytes are more or less independent of surface water because of the development of long tap roots reaching to deeper sources of soil water. This is especially true of species which grow on sand dunes such as the mesquite, whose roots may reach well over thirty feet beneath the surface. Roots, both lateral and tap, give firm anchorage against the action of strong winds; the numerous laterals protect against removal of soil from around the base of the plant.

The seemingly top-heavy tree yuccas maintain their upright position, in a land often visited by almost gale-strength winds, by having a resilient trunk and hundreds of long, pencil-sized anchoring rootlets striking almost directly outward from the fringes of their bulging bases. This is also true of the desert palm. Other desert plants such as the sand-mat or rattlesnake weed (*Euphorbia*) protect themselves from the almost constant wind by hugging close to the sand. Many fragile, weak-stemmed annuals grow up through the twiggy stems of rigid shrubs and gain protection against both wind and grazing animals.

Two main problems solved by desert plants also must be met by animals invading a desert environment: getting and preserving vital moisture; and securing adequate protection against excessive heat, sand and dust storms, cool to cold nights, and a host of special enemies.

Unlike plants, which must "stay put" and adapt themselves to weather changes or perish, animals can move about, go underground, or migrate from areas of poor food or water supply to more favorable spots. Thus animals are much more independent of their environment and can to a much greater extent "choose" their preferred habitat.

To conserve water, many animal dwellers venture forth to feed only in the cool of the night. This helps to explain why the desert appears so lifeless to the casual observer who goes abroad only during the day. Animals that forage over considerable reaches of wasteland usually rely for water upon the few scattered springs which they visit at dusk or during the night. Some of the smaller animals that seldom range farther than a few hundred yards from their home secure their water from the plants or animals upon which they feed. Several of the desert rodents, such as the desert hare and kangaroo rat, and many of the wild mice are capable of manufacturing water from their dry food. Such "metabolic water" may enable them to go through life without taking a drink. As a means of further conserving water some desert creatures such as the lizards and snakes void no liquid waste. Both feces and urine are voided in almost solid form.

Most of the desert animals spend much of their period of inactivity beneath the ground, where it is considerably cooler and somewhat moist throughout the year. The heat of the desert sun, even in midsummer, rarely penetrates more than a few inches below the soil surface, so that underground burrows are many degrees cooler than the air above them. Animals which excavate underground tunnels, such as the kangaroo rat, antelope ground squirrel, pack rat, and shovel-nosed snake, are called fossorial animals. Even the clumsy-appearing desert tortoise digs extensive subterranean chambers and burrows, both for places of hibernation and for shelter from the summer's fierce sun.

Several desert dwellers hibernate or go into a temporary torpor underground during the colder winter months. There is then both an inadequate food supply and a temperature so low that it keeps their bodies from normal functioning. This is especially true of the snakes and most of the lizards. In the case of the ground squirrels and bats (which resort to caves and rock crevices) and that peculiar hibernating bird, the poorwill, lack of food is probably the deciding cause of winter torpidity.

Underground retreats also afford protection from the sand and fine dust raised by the frequent windstorms. As a further protection against sand and dust, Nature has provided many of the desert creatures with smaller than usual ear openings, with the added protection of long hairs or scales partially covering them. The eyelashes may be longer than is normal, and the eyelids thickened. Even the nostrils in some species are provided with valves or can otherwise be tightly closed against sand and dust carried by gusts of wind.

Nearly all desert inhabitants are much lighter colored than

their near relatives living in moister climates. This has been cited as an example of a kind of "protective coloration" which enables these creatures to be overlooked by predators. It is assumed that this light color has been evolved over a long period of time. But just how their color change came about is still a problem to be solved by biologists. It is possible, they say, that a paler-colored body reflects more heat and aids in conserving moisture. Only incidentally perhaps does it camouflage the animal against detection by its enemies.

Essentially the same problems that face wild animals had to be solved by the various Indian tribes that made their homes in deserts. The majority of these peoples probably did not come originally into the desert expecting to make it their final home, but were pushed into it by stronger or more warlike tribes on the fringes of the wasteland. However, once they had made an adjustment to the new and severe conditions, they found the desert not only far from inhospitable but a place offering many advantages, such as a warm healthful climate, a ready supply of dry wood for their fires, and numerous fibers and other materials for the making of clothing and shelters.

The first great problem confronting these primitive peoples was that of water. Most of the Indians moved where they could live near lakes, water holes, springs, or some of the few perennial streams. Some of the tribes became more or less sedentary and were able to practice limited agriculture. As they were able to grow more food, they did not have to spend so much time in hunting and therefore had more leisure. With leisure time, it was possible to develop more elaborate rituals and costumes and to become what we term more "civilized." The Hopi pueblo Indians of mid-Arizona, as also their relatives in New Mexico, perhaps reached the highest development of all Indians of the American deserts. The tribes along the Colorado River, such as the Mohave and Yuma, were in some respects even superior to the Hopis. The Navaho and the Apache, coming later into the area, and changing from purely nomadic hunters and raiders into sedentary farmers, have recently become quite civilized and have been able to develop a comparatively high culture.

The most interesting Indian tribes of all are not the sedentary agriculturalists but the food-gatherers and hunters who braved the heat of the desert and not only survived but in some instances even thrived. Such tribes were the Cahuilla of California and the Pima and the Papago of Arizona. These peoples deserve our special respect and admiration as we learn how they experimented with an unusual number of living things of the desert, and found uses for most of them. Predominantly they were nomadic; they had no permanent communities or villages but lived apart as family units, coming together from time to time only for communal hunts or ceremonies. Because they were ever on the move, migrating from one portion of the desert to another as different food plants came into fruit or seed, they could carry only bare essentials and had little time for ornamentation or decoration. Basket making was perhaps their only truly artistic outlet, and in this art these desert tribes excelled. One is sometimes apt to look down upon such people because they were mere food-gatherers and basketmakers. Instead, these tribes should be admired for their persistence and ingenuity in solving the major problems of shelter and subsistence in a thirsty land. There were of necessity few lazy Indians in those bygone days.

The Cahuilla Indians of southeastern California solved the water problem, at least in some areas, by digging terraced wells in the sand dunes, thus enabling their women to walk to the water level to fill their ollas. Among the staple foods of the desert Cahuilla was the bean of the mesquite (*Prosopis*, 218*) and the seeds of the small annual sage called chia. In the higher elevations they utilized for food the young shoots of the agave or century plant (*Agave deserti*). These they roasted in stone-lined pits. Farther up the desert mountains they harvested the nuts of the piñon or nut pine (*Pinus monophylla*) and gathered acorns, which to make edible they crushed and leached with warm water to remove the poisonous and bitter tannic acid.

The Pima and Papago Indians of southern Arizona and adjacent Mexico depended much upon the flowers and fruit of the sahuaro cactus (*Cereus giganteus*, 159) as a source of food. They also roasted agave shoots and gathered mesquite pods. In many other ways they developed a material culture similar to that of the Cahuilla.

* The number refers to the illustration of the species in the section following Chapter 14.

2

WEATHER AND CLIMATE*

The unique features of the landscape that impress every visitor to deserts, wherever they occur, are the distinctive, sparsely distributed wild plants, the seas of rolling sand dunes, the barren rocky or pebbly surfaces, the dry stream beds, the irrigated oases. All of them owe their character to the desert climate and, above all, to the phenomenon of dryness. Furthermore, the principal broad differences found in deserts are the result of climatic differences, especially in the amount of rainfall and the degree of heat and cold.

The driest part of the North American deserts, with an average of less than five inches of rainfall per year, lies along the great trough of the Gulf of California, thence northward to Death Valley and westward across Baja California to the Pacific Ocean (Fig. 1). The driest spots have an average of less than two inches of rainfall per year, but even this minute amount is undependable. Several weather stations have recorded no rainfall at all for a year or more. Probably the longest drought ever recorded in North America lasted for three years, from February 1917 to January 1920, at Bagdad, in the Mohave Desert. During that period only one month had measurable rainfall—March 1919, with one one-hundredth of an inch. The fact that Bagdad has an average rainfall of 2.28 inches means very little when we know that some years had no rainfall and one year had 9.9 inches. Similarly Yuma, with a mean annual rainfall of 3.39 inches, ranged all the way from 11.41 inches in 1905 to .31 inch in 1953. Such irregularity is characteristic of deserts the world over.

Outward from the core or center of the driest desert areas, precipitation increases in all directions: very rapidly where high mountains are close at hand (westward in California and Nevada, eastward in Sonora), but only very gradually to the north and

* By Peveril Meigs, Ph.D.

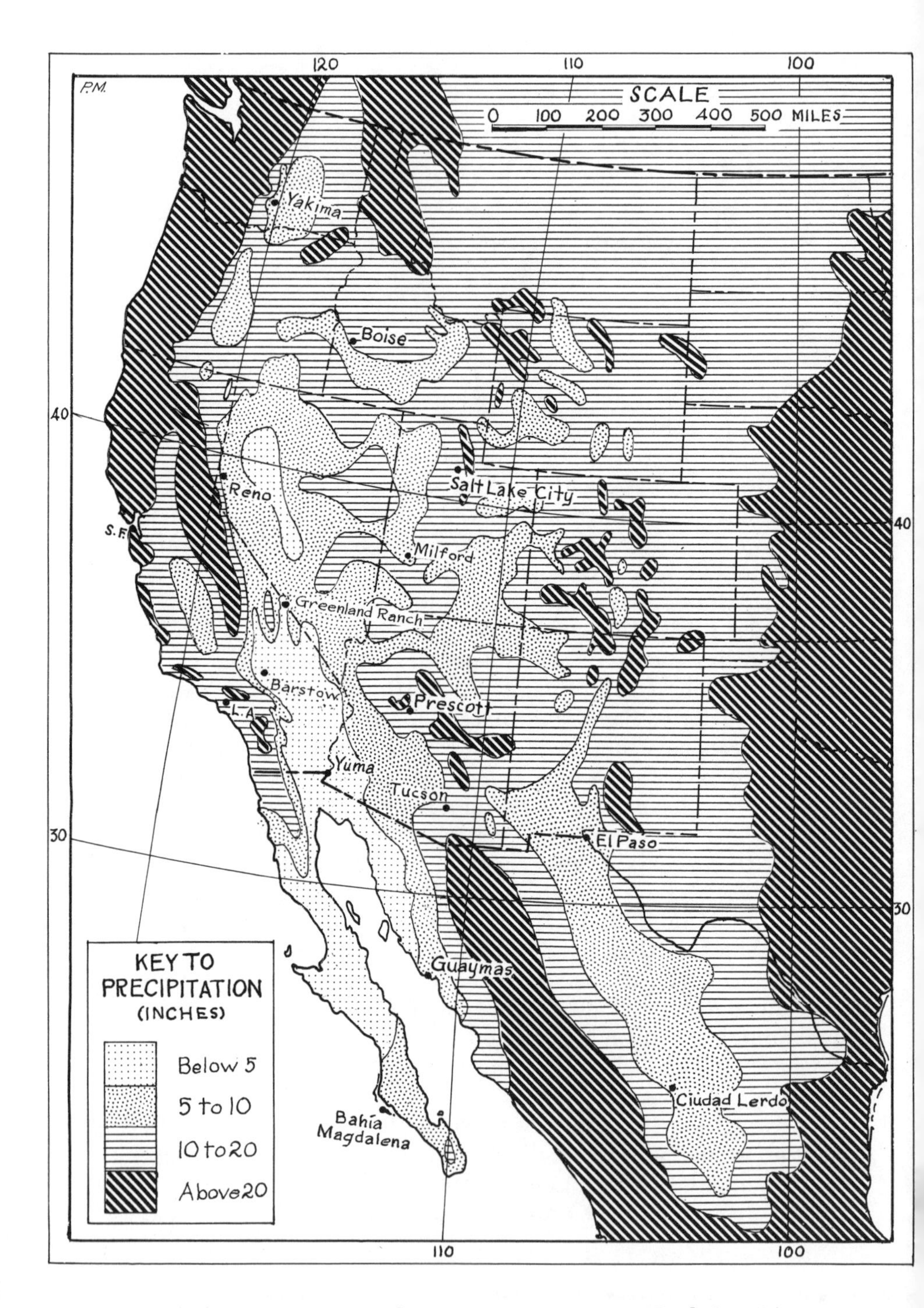

Figure 1. Average yearly precipitation in western North America.

east in the United States. The transitional area of five to ten inches of precipitation per year includes large portions of the Great Basin and the Chihuahuan Desert.

At its northern and eastern margins, the drier desert merges into semiarid steppe areas: upward into the higher plateaus and mountains; eastward in Texas it passes into the Great Plains. The semiarid lands of Utah, Arizona, and New Mexico in turn give way to the humid forests still higher on the mountains. Figure 1 shows only the amounts of precipitation: no attempt has been made to draw on it the precise limits between desert, semiarid, and humid areas. The drawing of such lines involves a consideration not only of the amount of precipitation but also of the temperature at the time of year at which rain falls. Where temperatures are low, moisture evaporates less rapidly than where they are high and is therefore available a longer time for the use of plants. An area along the Gulf of California might be a desert with twelve inches of rain, while an area of northern Washington with equal rainfall might be too wet to be a desert. Climatically speaking, Boise, Salt Lake City, and Prescott are semiarid rather than truly desertic.

The seasons in which rain occurs differ widely in different parts of our deserts. Winter precipitation, from storms sweeping in from the Pacific, is the rule in the western deserts of the three Pacific Coast states, western Nevada, and northern Baja California; summer rainfall is rare and slight (see Reno, Fig. 2). In the eastern and southern portions of the Mexican deserts, on the other hand, summer is the rainy season, with thunderstorms loosing the moisture that blows in from the Gulf of Mexico and Gulf of California (see Guaymas, Fig. 2). Between these two extremes is every gradation of seasonal distribution of precipitation. Occasional summer thunderstorms penetrate all the desert areas except the arid San Joaquin Valley of California, and winter storms have at times reached to the farthest bounds of the deserts in Mexico. The Rocky Mountains have a fairly even balance between winter and summer precipitation, and they form the dividing line between the western areas, having most precipitation in winter, and the eastern areas with summer rain predominating. From southwestern Colorado the division line bends southwestward across Arizona and the northwestern quarter of Sonora to the Gulf of California, leaving all of New Mexico, southeastern Arizona, and most of Sonora and Chihuahua on the summer-rain side.

The winter rains tend to fall gently and steadily for several

Table 1. AVERAGE PRECIPITATION (inches)

	YRS. REC.	J	F	M	A	M	J	J	A	S	O	N	D	ANNL.
Boise, Idaho	88	1.73	1.48	1.35	1.18	1.43	.92	.24	.19	.53	1.24	1.28	1.57	13.14
Green River, Wyo.	45	.42	.51	.58	.92	1.06	.65	.50	.74	.78	.96	.47	.37	7.96
Salt Lake City, U.	77	1.35	1.45	1.94	1.97	1.88	.87	.55	.84	.88	1.53	1.44	1.37	16.07
Reno, Nev.	81	1.40	1.09	.74	.45	.51	.31	.22	.21	.22	.38	.61	.94	7.08
Greenland Ranch, Calif.	42	.25	.31	.18	.12	.07	.02	.08	.12	.11	.10	.14	.19	1.69
Las Vegas, Nev.	14	.58	.46	.52	.27	.09	.02	.41	.39	.44	.27	.19	.49	4.13
Barstow, Calif.	40	.82	.65	.72	.18	.10	.11	.18	.27	.20	.38	.28	.61	4.50
Yuma, Ariz.	79	.39	.41	.32	.09	.03	.01	.19	.57	.40	.27	.23	.47	3.38
Guaymas, Sonora	..	.31	.24	.20	.12	.12	.04	1.85	2.99	2.13	.39	.43	1.14	9.96
Tucson, Ariz. (AP)	10	.72	1.01	.61	.31	.21	.16	1.98	2.17	1.46	.63	.45	.82	10.53
San Ignacio, B. Cfa.	3	.04	.16	.20	0	0	0	.55	.43	.35	.08	.12	.63	2.56
La Paz, B. Cfa.	..	.12	.43	.04	T	T	T	.24	1.65	2.05	.39	.51	1.34	6.77
Prescott, Ariz.	52	1.80	2.20	1.56	1.20	.44	.34	2.62	3.39	1.98	.99	1.08	2.38	19.98
El Paso, Texas	72	.46	.40	.32	.27	.35	.63	1.71	1.56	1.21	.85	.48	.52	8.76
Ciudad Lerdo, Dur.	..	.24	.16	.12	.12	.63	1.06	1.73	1.26	2.68	.75	.39	.47	9.61

NOTE: 0 = none, T = trace.

hours or days at a time. The summer rains are usually concentrated in brief thunderstorms lasting from a few minutes to several hours. Often the summer storms yield little rainfall, since the rain evaporates before it reaches the ground. But occasionally a terrific cloudburst swells the dry arroyos into raging torrents that accomplish more erosion in a few hours than normally occurs in the course of several years. The wise traveler does not camp in an arroyo bed near the mountains when storm clouds are in evidence, especially in summer. But actual probability of occurrence of rain

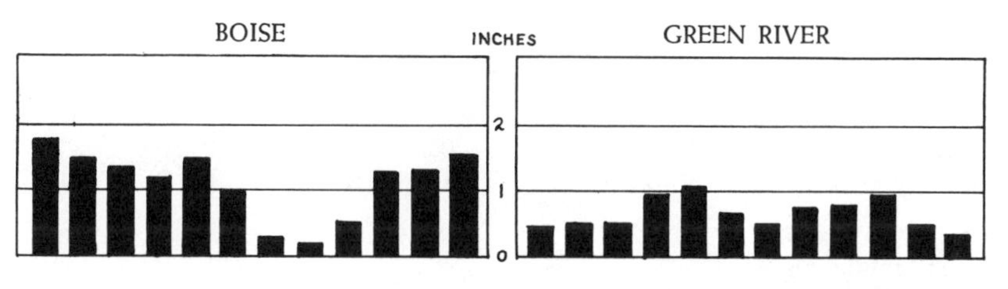

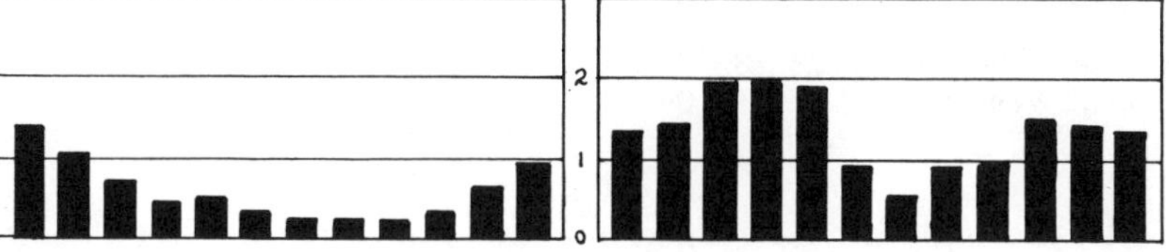

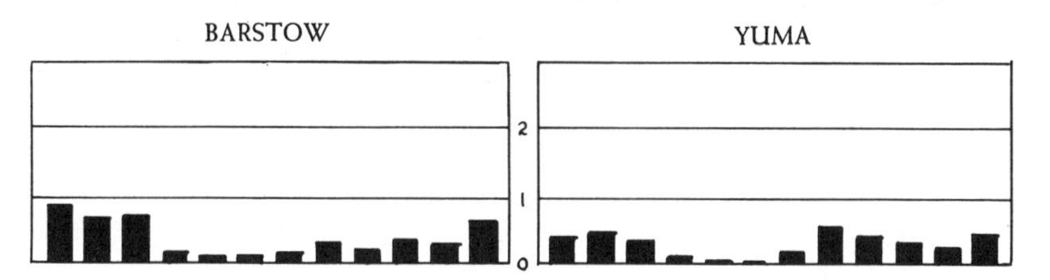

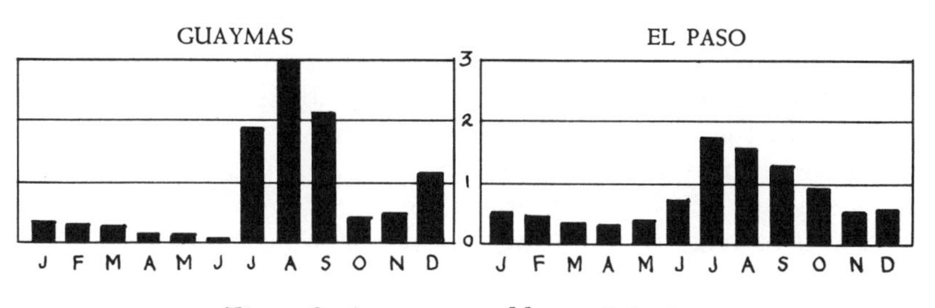

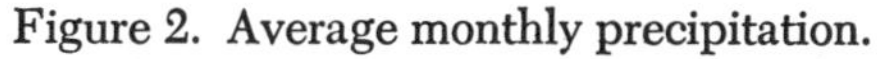

Figure 2. Average monthly precipitation.

in a given week is slight, especially in the summer in the most western deserts, where the probability approaches zero (Table 2) and the camper can leave his tent at home.

Table 2. AVERAGE NUMBER OF DAYS WITH PRECIPITATION

	YRS. REC.	J	F	M	A	M	J	J	A	S	O	N	D	ANNL.
Boise, Idaho	11	11	11	9	8	8	7	2	2	3	7	11	11	90
Salt Lake City, U.	77	10	10	10	9	8	5	4	5	5	7	7	11	91
Reno, Nev.	63	7	6	6	4	4	3	2	2	2	3	4	6	49
Las Vegas, Nev.	14	3	3	3	2	2	*	2	2	2	2	2	3	26
Yuma, Ariz.	73	2	2	2	1	*	*	1	2	1	1	1	2	15
Guaymas, Sonora	..	1	1	1	0	0	0	5	6	3	1	1	4	23
Tucson, Ariz.	10	4	4	4	2	1	1	9	9	5	4	2	4	49
San Ignacio, B. Cfa.	3	1	2	1	0	0	0	2	4	2	2	3	4	21
Bahía Magdalena, B. Cfa.	4	4	2	4	1	0	0	1	2	6	1	6	5	26
Prescott, Ariz.	8	5	6	6	4	1	1	11	10	5	4	3	6	62
El Paso, Texas	72	3	3	3	2	2	4	8	8	6	4	3	4	50
Ciudad Lerdo, Dur.	..	1	0	1	1	2	3	5	4	6	4	2	4	33

* Less than ½.

In the higher, colder deserts and steppes of the Great Basin and Navahoan uplands, there is considerable winter snowfall. Farther south there is a great decrease in snowfall; for instance, El Paso 2½ inches, Las Vegas 2 inches, and Tucson 1 inch. Even Yuma has been known to have a trace of snow in December and January. Most of the Sonoran and Chihuahuan deserts are snowless, though snow at times makes the near-by "tentlike mountains gleam like the encampment of some mighty host." On all the high deserts, even including the upper Mohave, snowfall is likely to occur, at least on the higher mountains, any time between mid-November and mid-April. From December to March, the eastern side of the Great Basin can expect a fall of about one foot of snow per month; the western side, about six inches.

There are profound temperature as well as precipitation contrasts among the North American deserts. We are apt to think of deserts as places of burning sun and scorching heat, but days of pitiless sun in a cloudless sky are normal only for a desert summer. The core of our own desert realm yields second place to none in the intensity of its summer heat. The temperature of

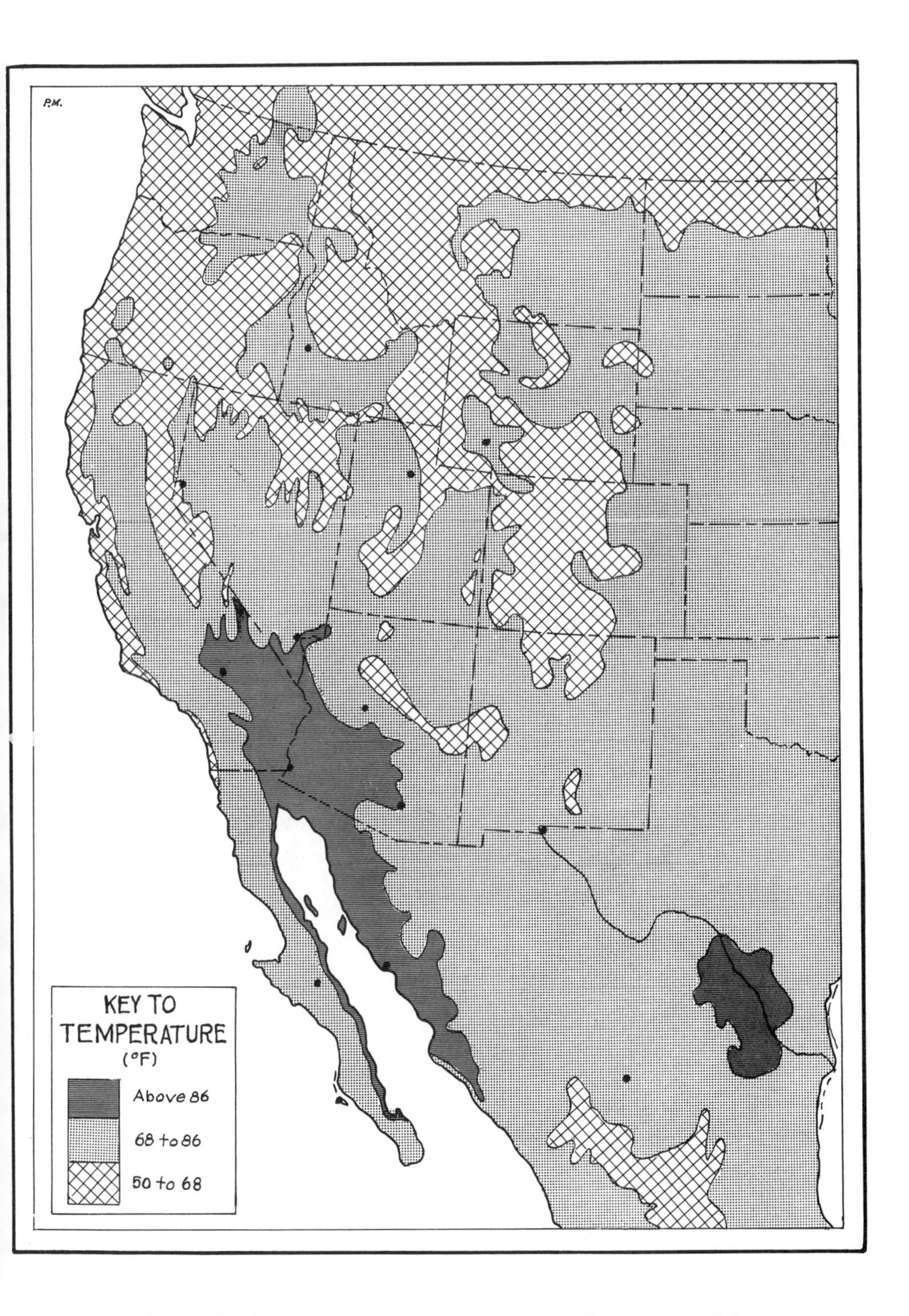

Figure 3. Average temperature in western North America in July.

Table 3. MEAN TEMPERATURE (°F), LATITUDE, AND ELEVATION

	N. LAT.	ELEV. (*ft.*)	YRS. REC.	J	F	M	A	M	J	J	A	S	O	N	D	ANNL.
GREAT BASIN DESERT																
Boise, Idaho	43°34′	2,842	38	28	34	41	49	56	64	72	71	61	50	40	30	50
Green River, Wyoming	41°32′	6,109	46	18	23	33	43	52	64	69	67	57	54	32	21	43
Salt Lake City, Utah	40°46′	4,260	77	29	34	42	50	58	68	77	75	65	53	41	32	52
Reno, Nevada	39°30′	4,397	63	32	36	41	49	55	63	71	69	61	51	44	34	50
MOHAVE DESERT																
Greenland Ranch, California	36°28′	–178	41	52	58	66	75	84	94	102	99	89	75	61	52	76
Las Vegas, Nevada	36°10′	2,006	42	45	50	57	65	72	80	86	84	78	66	54	46	65
Barstow, California	34°54′	2,105	32	46	50	56	62	69	77	86	83	75	64	54	46	64
SONORAN DESERT																
Yuma, Arizona	32°44′	138	71	55	59	64	70	77	85	91	90	85	73	62	56	72
Guaymas, Sonora	27°56′	10	16	64	66	69	73	78	84	87	87	86	81	73	65	76
Tucson (airport), Ariz.	32°08′	2,558	10	50	54	58	66	74	82	87	84	81	70	59	52	68
VIZCAINO-MAGDALENA DESERT																
San Ignacio, B. Cfa.	28°56′	*c.*500	3	58	60	62	67	72	76	83	81	80	73	67	63	70
Bahía Magdalena, B. Cfa.	24°38′	*c.*0	4	69	68	69	68	69	71	77	80	82	79	76	72	73
NAVAHOAN DESERT																
Prescott, Arizona	34°33′	5,354	52	34	38	43	51	59	67	73	71	65	54	43	36	53
CHIHUAHUAN DESERT																
El Paso, Texas	31°48′	3,920	71	45	50	56	64	72	81	82	80	75	65	52	45	64
Ciudad Lerdo, Durango	25°32′	3,740	..	57	61	67	73	79	81	80	80	76	71	63	56	70

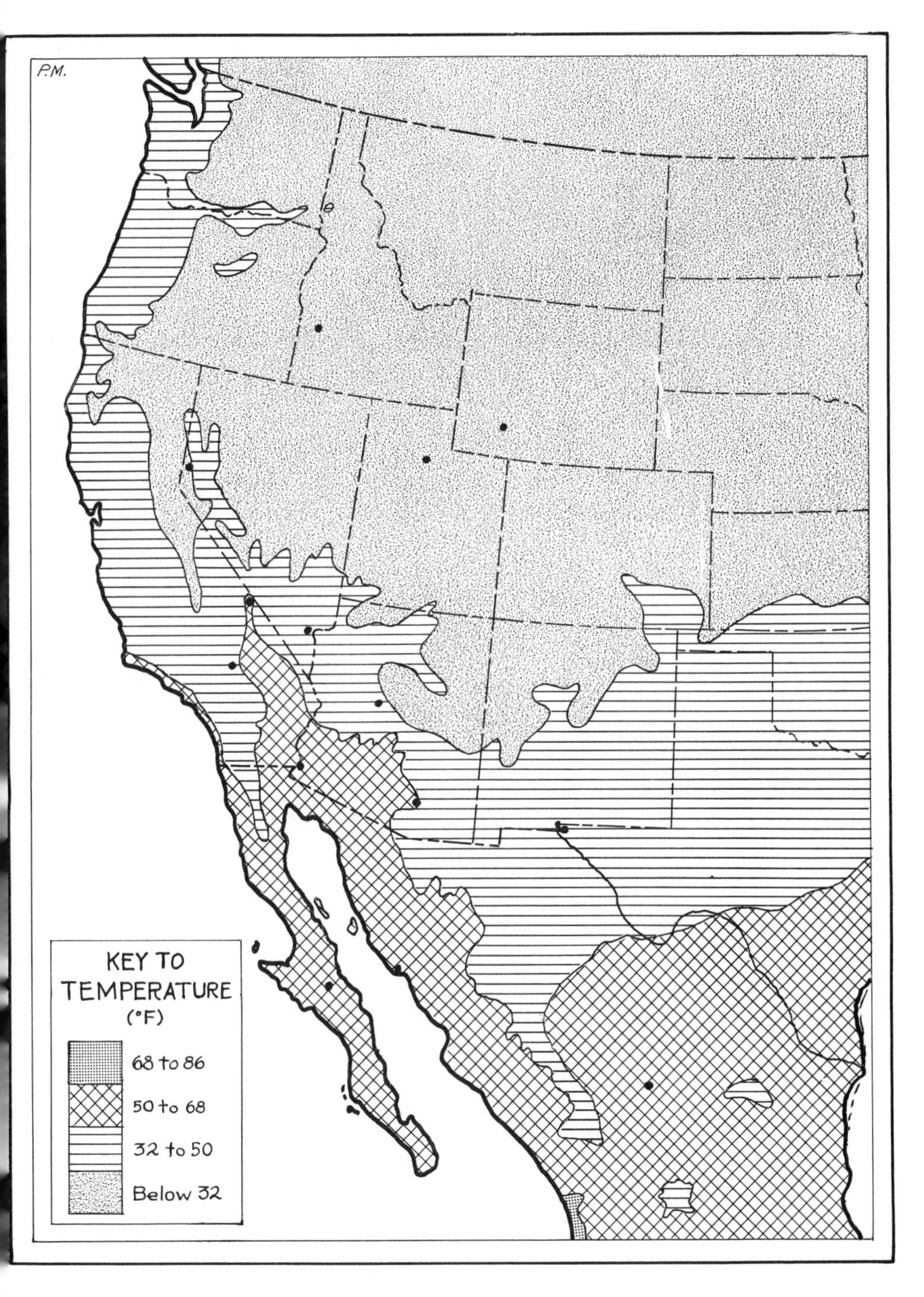

Figure 4. Average temperature in western North America in January.

134° F. reached on July 10, 1913, at Greenland Ranch, Death Valley, has not been surpassed at any standard observation station in the world, with the possible exception of Azizia, Libya, whose record of 136.4° F. has not been universally accepted by meteorologists. Death Valley is ordinarily much hotter than Azizia: the average daily maximum temperature at Greenland Ranch is 116° in July, compared with 99° at Azizia. To drop into Death Valley at Greenland Ranch, 178 feet below sea level, on a summer afternoon is almost like entering a furnace.

The distribution of the very hot summer desert is shown by dark shading on Figure 3. Where the average temperature is above 86°, the thermometer goes above 100° nearly every afternoon of the month (Table 4). Even in the northern and elevated

Table 4. MEAN DAILY MAXIMUM TEMPERATURE (°F)

	YRS. REC.	J	F	M	A	M	J	J	A	S	O	N	D
Boise	11	35	44	52	62	71	77	90	88	77	65	48	40
Salt Lake City	77	36	42	51	60	69	80	89	87	77	64	50	39
Reno	63	43	48	54	62	69	79	88	87	78	67	55	45
Greenland Ranch	35	66	72	81	90	99	109	116	114	106	91	76	66
Las Vegas	14	57	62	70	81	90	99	106	104	97	83	69	60
Yuma	71	67	72	78	86	93	102	106	104	100	88	76	67
Guaymas	..	74	76	78	82	87	92	94	95	93	89	82	74
Bahía Magdalena	4	76	76	77	77	76	78	83	87	88	86	83	80
Prescott	8	49	54	59	69	77	85	90	88	84	73	61	52
El Paso	64	57	62	69	77	86	94	93	91	86	78	66	57
Ciudad Lerdo	..	70	76	81	87	91	90	90	86	86	82	75	68

deserts the afternoon temperature usually reaches 90° in July. Fortunately for human comfort three factors help make the summers bearable. First, the relative humidity is very low during the heat of the day; second, the breeze is usually strongest in the afternoon; and, third, nights cool off rapidly. In the Colorado Desert, temperatures at night drop about 30° below the afternoon peak, so that a temperature of 105° might be expected to drop to about 75° by the following morning (Tables 4 and 5). When one sleeps in the open, radiation from the body to the open sky causes a further feeling of coolness: even with air temperature around 80° a light blanket may feel comfortable outdoors. Where the desert reaches the sea along the coasts of Sonora and Baja California the daily swing of temperature is only about half as great as farther inland. In all of our North American deserts there are places where at night a light sweater is comfortable, even in summer.

Table 5. MEAN DAILY MINIMUM TEMPERATURE (°F)

	YRS. REC.	J	F	M	A	M	J	J	A	S	O	N	D
Boise	11	19	27	31	38	44	50	58	56	48	40	31	25
Salt Lake City	77	22	26	33	40	48	56	64	63	53	43	32	25
Reno	63	20	25	29	34	40	47	53	51	44	35	28	22
Greenland Ranch	35	38	44	51	60	69	78	87	84	73	59	46	39
Las Vegas	14	29	34	40	49	57	65	71	70	62	50	36	32
Yuma	71	42	46	50	54	60	68	77	77	70	58	48	43
Guaymas	..	56	58	59	64	69	77	81	80	78	72	65	57
Bahía Magdalena	4	61	60	60	60	62	65	70	74	74	72	69	65
Prescott	8	22	26	30	38	45	52	60	59	53	41	29	25
El Paso	71	32	37	42	50	58	67	70	68	63	52	39	33
Ciudad Lerdo	..	43	47	52	58	66	68	68	68	65	58	50	45

The actual changes in weather from hour to hour on a typical summer day in the vicinity of Yuma are shown in Figure 5. The sharp increase in heat starting about sunrise, the midafternoon peak, the vastly greater rise of soil temperature, the steady drop from the afternoon peak to dawn, the drop of relative humidity to 12 per cent in the heat of the day and its rise to 60 per cent at dawn, and the increase of wind from a calm at night to a fresh breeze toward evening—all are characteristic of the desert. The exact values, of course, differ greatly from day to day and place to place.

The efficiency of dry air in cooling the body is suggested by the enormous amount of evaporation in the desert. On the mesa a few miles south of Yuma a standard Weather Bureau evaporation tank shows an average evaporation of 115 inches per year during 32 years of record. At Bartlett Dam, 35 miles northeast of Phoenix, annual evaporation totals 122 inches, ranging from a monthly average of 4 inches in January to 17 inches in June and in July. The stronger winds at Bartlett Dam account for the greater evaporation than at Yuma. Boulder City, Nevada, has about 120 inches of evaporation. In the Great Basin, evaporation in the summer months averages about 10 inches per month, compared with 15 to 17 inches in the Colorado and Arizona Upland deserts.

The cooling effect of evaporation upon man involves the expenditure of great quantities of sweat. It has been estimated that a man of average weight walking eight hours per day in the Colorado Desert in July should drink about three gallons of liquid per day if he is to retain his weight. If he sits in the shade all day he can get along on six quarts. In the cooler deserts of the Great Basin these quantities can be cut in half. He can drink much less

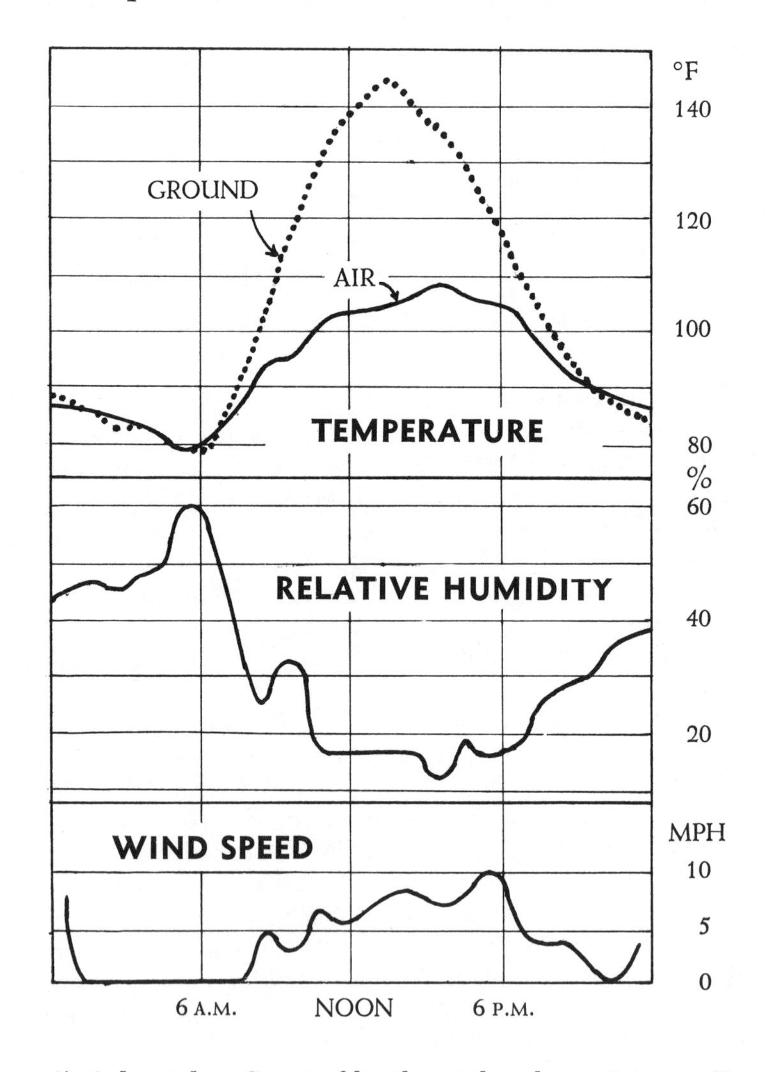

Figure 5. A desert day. Curves of hourly weather observations near Yuma, Arizona, August 5, 1951, a typical day in the Sonoran Desert.

than this for several days or even weeks without suffering, but he will lose weight by dehydration during this period. Working in the desert, one soon learns to take all possible advantage of the cooler hours. It is desirable to have the camp cooking and dish-washing out of the way before things start to heat up at sunrise. Experienced *vaqueros* make it a practice to be in the saddle by dawn. If by some unlucky fate a person is stranded in the desert without water, far from a water source, the worst thing for him to do is to strike out across country during the heat of the day. He might not last even one day using these tactics. The best course is to sit quietly in the shade during the day and walk only at night. In this way, it is estimated that a man without water can walk about two nights or 36 miles in the Colorado Desert, and survive for two or three additional days by resting. In the Great Basin and Painted deserts these times might be doubled.

Summer lasts only two or three months in the Great Basin and Painted deserts, about six months in the Mohave, and seven to nine months in the Sonoran Desert. It is preceded by a fresh, delightful, often all too short spring period of slowly rising temperature, and followed by a shorter fall period of rapidly falling temperature.

Winters run the gamut from very cold to warm. In the Great Basin and in the higher uplands of other deserts, temperatures may fall below freezing many nights from December through March. Even subzero temperatures occur once in a while (Table 6) in the cold-winter deserts. In the Sonoran Desert winter nights are chilly but seldom drop to freezing. Days are pleasantly clear and sunny. Only along the coasts of the Gulf of California and Pacific Ocean are frosts unknown: the marine and subtropical influences there reduces the annual as well as the daily extremes.

In all the deserts woolen clothing is comfortable in winter, at least at night. In the cold-winter deserts, overcoats or equivalent protection are needed from about November to March. Nowhere except in the southern part of the deserts bordering the Gulf of California is it safe in summer to sleep in the open without having available a blanket or sleeping bag. Even between latitudes 27° and 28° N. on the Pacific side of Baja California, a traveler visiting a local gold rush in the low Sierra Magdalena near the coast in summer found many of the miners suffering from rheumatism and other ailments brought on by cool windy weather and inadequate protection, and advised that heavy blankets were needed at night.

The length of the growing season, or the time between killing

Table 6. RECORD HIGHEST AND LOWEST TEMPERATURES (°F)

	YRS. REC.	J	F	M	A	M	J	J	A	S	O	N	D
Salt Lake City	77	62	68	78	85	93	103	105	102	97	88	74	68
(to 1950)		–20	–13	0	18	25	32	43	42	29	22	–2	–10
Las Vegas ..	14	76	80	89	99	108	116	117	116	113	100	84	78
(to 1950)		8	17	19	31	38	48	56	54	43	32	15	14
Yuma	73	84	92	100	107	120	119	120	119	117	108	96	83
(to 1952)		22	25	31	38	39	50	61	58	50	38	29	22
Bahía Magdalena	4	90	86	88	84	89	93	93	96	97	93	93	90
(to 1940)		52	46	52	55	54	54	61	66	70	66	57	49
El Paso ..	64, 71	77	86	93	95	102	106	105	103	100	94	85	77
(to 1950)		–6	5	14	26	36	46	56	52	41	26	11	–5
Ciudad Lerdo	..	88	92	96	100	103	105	100	100	98	93	91	89
		18	27	28	35	45	58	60	57	49	41	36	23

frosts, varies greatly. At Rawlins, Wyoming, at the edge of the Red Desert 6,744 feet above sea level, the growing season averages 97 days in length; in the Baja California and Sonora coastal deserts frosts are unknown. The following tabulation gives the average dates of last and first freezes of spring and fall, and the length of the growing seasons. In individual years, the dates may be as much as a month or more earlier or later.

	AVERAGE DATES OF KILLING FROST		AVERAGE LENGTH OF GROWING SEASON (DAYS)
	Last in spring	First in fall	
Rawlins, Wyoming	June 3	Sept. 8	97
Reno, Nevada	May 16	Sept. 30	137
Boise, Idaho	April 28	Oct. 22	177
Salt Lake City, Utah	April 19	Oct. 18	182
San Marcial, N.M. (upper Rio Grande valley)	April 8	Oct. 22	197
El Paso, Texas	March 20	Nov. 10	235
Parker, Arizona	March 8	Nov. 22	259
Yuma, Arizona (valley) ..	March 6	Nov. 21	260
Yuma, Arizona (mesa) ...	......	......	365
Guaymas, Sonora	......	......	365

Differences in elevation account for many local temperature differences. Over the world as a whole, temperature drops about 1° F. for each 300-foot rise in elevation. This figure is subject to infinite variation in detail. In summer in the daytime the rate of cooling is greater than the figure given. On still, clear nights, on the other hand, mountain slopes are actually warmer than the basins or valleys at their feet. This inversion of the "normal" situation occurs when cold air drains downslope. Systematic measurement made at the former Desert Laboratory near Tucson and in a near-by valley 335 feet lower showed the minimum night temperatures to be consistently lower in the valley than at the laboratory. The difference ranged from 8° or 9° F. during the relatively humid midsummer to 17.8° F. in May. One night the difference amounted to 24° F. At elevations of 3,000 to 4,000 feet on the near-by Santa Catalina Mountains, spring opens earlier than at the Laboratory (elevation 2,663 feet). The existence of relatively warm zones on slopes above valley floors is a common feature of deserts, where rapid radiation from the dry ground results in greater temperature contrasts than in humid regions. The "thermal belts" are sufficiently marked to affect the type of vegetation growing in them.

Sandstorms and dust storms are among the most spectacular features of our American deserts, as also of most deserts the world over. On the average the winds of the southwestern deserts are no stronger than those of the eastern United States. The occasional strong desert winds are more conspicuous because the loose dry soil, unprotected by a continuous mantle of vegetation, can be picked up readily. In order for a sand or dust storm to develop, there must be loose material available and sufficiently strong winds. The sand deposited in arroyo beds is the primary source of much sand, and muds spread out by occasional floods are the source of much of the dust in a storm. Disturbance of the ground surface by traffic, cultivation, or other agencies markedly encourages dust storms. Where all the loose material has been swept away by previous storms, leaving only very coarse gravel and rocks on the ground, the wind will be unable to raise sand or dust however strongly it blows. Other conditions being favorable, sand and dust begin to move when the wind reaches a speed of about 15 miles per hour. At 20 miles per hour a mild dust storm may develop; at 30 miles per hour a severe sandstorm may develop. A gusty wind is more favorable for the production of sandstorms than a steady wind of the same average speed.

Spring is the windiest time of year and then is there greatest likelihood of sand and dust storms. Even in April, the windiest month of the western deserts, the average wind speed is only 6.6

miles per hour at Yuma and 11.6 in El Paso. These are average figures based on the entire day; stronger winds are usual in the afternoon. The fastest mile of wind ever recorded at Yuma attained a speed of 56 miles per hour; at El Paso, 70. These maximum winds were not grouped into any particular season; they were scattered throughout the year. On the average, however, winter has the lightest winds. The December average at Yuma is 5.5 miles per hour; at El Paso, 8.5.

A well-developed sand and dust storm is a memorable event. The sun is obscured, visibility may drop to zero, and movement along a highway becomes impossible. Fierce blasts of sand "frost" the windshield until it is useless, and cut away the enamel from the body of the car. In the Mohave and Colorado deserts it has sometimes been necessary to bar all traffic from highways for a period lasting from several hours to a day or two, and sand removal is a recurrent problem of desert roads even where only shallow drifting occurs. In the typical milder storm the sand drifts along the ground like a moving carpet, not more than knee high. Even this movement is sufficient in the course of time to carve and smooth the rocks, develop planed surfaces on the sides of stones facing the wind, and finally deposit the sand in the form of dunes.

Winds blow from nearly all directions as the storms with which they are associated move along. Two types of storms predominate: the North Pacific type and the tropical type. The Pacific storms move in from the west and south. Along with the northward and southward seasonal shifting of the position of the sun relative to the earth, the storm tracks shift too. In winter and spring the Pacific storms are in control. Many of them become dessicated as they cross the Sierra and Cascade mountain barriers, many others bypass the deserts to the north, but luckily some do bring precipitation to the deserts, and a few pass all the way across the Chihuahuan Desert or to the southern parts of the Magdalena Desert. In summer and fall, most of the Pacific storm tracks are near or north of the Canadian border. At this season the hot humid air from over the tropical waters of the Gulf of Mexico and Gulf of California brings heavy rains to the southern and eastern edges of the deserts and occasionally as far north and west as Yuma, Tucson, and Salt Lake City.

In winter a persistent oceanic high-pressure area off the coast of Baja California and southern California causes northwesterly winds to prevail alongshore and helps shield the deserts from the marine storms of that season. A special feature of the west coast deserts of Baja California, at least as far south as Vizcaíno Bay, is the sea fog that drifts over the land almost daily in spring and summer, resulting in a cool or mild desert with moist air, quite unlike the exceedingly dry deserts of the interior.

Storm tracks are modified or deflected also by pressure areas over the land that may persist for days or weeks at a time. In winter, a *high* is often present in the Great Basin. Pacific storms are then deflected to the north, or rarely to the south, leaving the Basin clear, cold, and relatively calm. Under these conditions an ice fog, known as *pogonip* (a Paiute Indian word), sometimes develops in central Nevada in December or January and lasts for several days. A fog of similar origin, but not frozen, is typical of the arid parts of the San Joaquin Valley in California.

In summer an oval area of intensely low pressure prevails in the western part of the deserts, with the axis of the *low* running approximately through the Chihuahuan Desert, Yuma, and the west side of the Great Basin. Southerly winds prevail to the south and east of the *low*, bringing to Yuma in July and August its most uncomfortable weather, a result of hot humid air from the Gulf of California. Under these conditions local thunderstorms are frequent. Brief dusty periods often herald the first windy blasts of these strange picturesque thunderstorms.

The worst storms of the entire desert region of North America, both for size and violence, are the chubascos (Spanish, "squalls") of the hot southern deserts—tropical hurricanes much dreaded by navigators in the late summer and fall. These great rotating storms, on the order of 50 to 100 miles across and with winds of 80 to 100 miles per hour, commonly start off the west coast of Mexico about 10° to 15° north of the equator. They move in a northwesterly track, many of them skirting the Mexican coast into the Gulf of California and disappearing inland over Sinaloa or Sonora. Very rarely, however, one of these great storms may travel as far as to the Colorado or even to the Mohave Desert before dying out. Others move to the west of Baja California, disappearing finally at sea.

Since very early times voyages from Mexican ports to Gulf ports of Baja California have been hampered by chubascos. The settlement of Baja California missions by the Dominicans, for example, was delayed in October 1772 when a vessel was wrecked en route from San Blas to Loreto, and Father Iriarte (president of the proposed Dominican missions of Baja California) and two other priests were drowned. Along the coast of the Mexican mainland the violent southerly winds on the east side of a chubasco are known as the Lash of St. Francis. Damage is done by tremendous cloudbursts as well as by wind. In September 1941, two chubascos with destructive floods struck the Cape region of Baja California. The first destroyed a substantial highway near La Paz. The second washed out an entire village at Cape San Lucas. As much rain fell in one day as is normal for the whole year. La Paz received 6.65 inches of rain on September 10, 1941, which compares with the average yearly total of 6.77 inches. San Jose del Cabo averages 12.06 inches per year, but on September 20, 1941, the chubasco brought it 13.39 inches! As in most deserts, rainfall averages have little meaning here.

Of the local winds of the desert, the whirlwind, dust devil, or tornillo, is the best known. It appears to be the product of the irregular upward rush of heated air on a calm day. It can be seen as well as felt: a slender spinning column of dust and debris from a score to several hundred feet high, swaying slightly as it moves across the desert with its narrow end concentrated on the ground, stirring up the dust that feeds the column. Often several of these spectral columns can be seen at the same time moving in silent procession. As a dust devil approaches, its dry swishing sound can be heard, and as you pass through it your eyes, ears, and nose are suddenly filled with a flurry of sand and dust. The wind is seldom strong enough to do damage and there is little inconvenience other than a momentary irritation. It is not surprising that these ghostly apparitions have been personified by some of the Indians of the desert.

Where there are long slopes and there is no strong general wind, the local wind often alternates its directions between night and day. At night a downslope or mountain breeze may develop, carrying scents of pine or fragrant shrubs and refreshing coolness to lower levels, while during the day an upslope or valley breeze ascends the mountains. These slope breezes, like any other wind, become stronger when channeled into a valley or pass. One of the most striking examples of such a concentrated wind (not a slope wind in this instance) occurs in San Gorgonio Pass, the link between the cool Pacific lowlands around Los Angeles and the heated Colorado Desert. During the windy days of spring when there are fogs along the coast and later when the greatest heat occurs, westerly winds often acquire sandstorm intensity as they traverse the gigantic funnel of the pass.

Sea breezes are a pleasant daily feature modifying to a certain extent the hot summers of many parts of the Gulf of California shores. Detailed information on their hours of occurrence, regularity, and distance of penetration are lacking.

Nowhere in the world is there as great a concentration of different types of desert climate as in western North America. It is interesting to know which of the principal desert areas of the world resemble particular deserts of North America. Of course no two areas have exactly the same climatic characteristics, but the following comparisons are reasonably close in regard to amount and seasons of rain and snow, coldness of winter, and hotness of summer.

The desert and steppes of the Great Basin, with cold, somewhat snowy winters and hot dry summers, find their closest parallel in the deserts and steppes of the western interior of Asia. The drier, warmer parts of Nevada and Utah are climatically analogous to the Russian deserts centering about the Aral Sea—the Kyzyl Kum and Kara Kum deserts and bordering foothills. Salt Lake City and Boise correspond to somewhat moister areas: the mountains of Iran and the Samarkand area, the Anatolian Plateau of Turkey, and the lower slopes of the Atlas Mountains of northwestern Africa. The still colder winter areas of Wyoming and Montana, without distinct seasonality of precipitation, as represented by Green River, resemble the Gobi Desert and the desert north of the Caspian Sea near the mouth of the Volga.

The Mohave Desert and San Joaquin Valley, with cool, relatively moist winters and hot dry summers (Barstow, Las Vegas), are climatically close to the Plateau of Iran and the northwestern part of the Algerian Sahara.

The Chihuahuan Desert has its analogues in the southern hemisphere. The northern part (El Paso), with cool winters and most but not all of its rain falling in summer, compares with the eastern part of the Karroo in South Africa and with the sheep-raising Riverina district of southeastern Australia, though the latter has a little more rain than El Paso. The southern part (Ciudad Lerdo), with milder winters and a sharp concentration of rain in the hot season, is like the Kalahari Desert and the west coast of Angola in southern Africa.

The interior of the Magdalena Desert (San Ignacio), in southern Baja California, is drier but otherwise quite similar to the southern part of the Chihuahuan Desert. The coast (Bahía Magdalena) is warmer in winter and moister than the interior, thus resembling the tropical coastal lands of Mauritania in French West Africa, the northern end of the coast of Peru, and the desert coast of Venezuela.

The Vizcaíno Desert, in central Baja California, is climatically typical of many of the well-known mild coastal deserts with winter rain: the Nile Delta, the desert coast of Libya and Tunis, the Negev region of Israel, the Casablanca coast, and the southern end of the Atacama Desert in Chile. On the basis of meager observations, we are led to believe that the immediate coast of the Vizcaíno Desert is bathed in fog more frequently than any of the above places.

The Arizona Upland (Tucson), with mild winters, hot summers, and a preponderance of summer rain, finds its climatic analogues in parts of the Kalahari, the higher portions of interior Australia, and the Mendoza oasis desert country of western Argentina.

The closest American parallel to the huge, hot deserts of the Old World is the low-lying core of the Sonoran Desert. Some of the driest parts of the Colorado Desert approach central Saharan conditions, though none are quite as dry as the heart of that yast expanse. The north coast of the Red Sea and the Great Australian Desert are closer analogues of our driest deserts. Most of the Colorado Desert, with mild winters, very hot summers, and rainfall inclined toward a winter maximum, as at Yuma, is very similar in climate to lower Iraq from about Baghdad to the Persian Gulf, the coast of the Persian Gulf in Iran and its extension into western Pakistan, the valley of the Jordan River in Israel, and the northern Sahara in southern Tunisia and adjacent Algeria. Yuma, however, has a higher summer humidity than the northern Sahara. The Sonoran Desert (Guaymas), sloping to the Gulf of California, differs from the Colorado Desert in having a marked concentration of rainfall in the hot season. It is very much like the Thar Desert of Pakistan and India, which has a brief, sharp monsoonal rainy season; also like the northern tropical part of the Australian deserts. The southern part of the Sahara is a little warmer in winter than the Sonora desert but in other respects is quite similar.

All in all, one who has an intimate knowledge of all parts of the North American deserts will have a good understanding of desert climates in most parts of the earth.

3

THE *Chihuahuan* DESERT

The large CHIHUAHUAN DESERT, named after the Mexican state in which much of it lies, extends farthest south of all North American deserts, even farther south than the desert areas near the tip of the peninsula of Baja California. It takes in the great triangular-shaped intermountain plateau of northern Mexico, including the broad tablelands comprising nearly two-thirds of the State of Chihuahua, much of the states of Coahuila, San Luis Potosí, and northwestern Hidalgo; included also in this desert are portions of western Texas, southern New Mexico, and southeastern Arizona (Wilcox Playa). The great Mexican plateau rises gradually in elevation from the Rio Grande southward. The Chihuahuan Desert ranks in size next to the Great Basin (Sagebrush) Desert.

Conspicuous on this desert are the rolling grasslands of the volcanic soils, the creosote bush plains, and the cactus savannas and agave thickets of the limestone soils. There are numerous bolsons (Spanish, "large purses"), a name given to large, dry inland basins which have no drainage outlet to the sea. In many of these bolsons are ephemeral shallow lakes, in which runoff waters from the encompassing slopes and mountains collect during the rainy season, only soon to evaporate and leave behind upon their flat beds thin layers of gray calcareous clays, alkali, or salt encrustations of almost snowy whiteness. Such deposits, often of unusual economic importance, have sometimes accumulated to a great depth, indicating filling and evaporation cycles over a very long period of arid conditions. The general drainage of the northeastern part of the Mexican portion of the Chihuahuan Desert, beyond the area of bolsons, is toward the Rio Grande.

Lying isolated or in parallel chains with general north-south or northwest-southeast orientation are many steep-walled and sometimes massive mountains which rise above the broad brush- and grass-covered, nearly level desert plains, like islands from the sea. Many of them, especially in eastern Chihuahua and Coahuila, are composed of or capped by Cretaceous limestone. Evidence of past volcanic activity is evident in many places, particularly in central and western Chihuahua.

The best plant indicators for northern parts of the Chihuahuan Desert are the creosote bush (*Larrea divaricata,* 238); the common and always conspicuous small yellow-flowered agave (*Agave lechuguilla,* 197) of limestone soils, commonly known as lechuguilla (Spanish for "little lettuce"); the narrow-leaved sotol (*Dasylirion wheeleri,* 164); the bisnaga or barrel cactus (*Echinocactus wislizenii*); the forbidding, spine-studded shrub called allthorn (*Koeberlinia spinosa,* 147); and the yellow-flowered varnish bush, known to botanists as *Flourensia cernua* (265) and to native Mexicans as "hojase"; often it is associated with creosote bush. The honey mesquite (*Prosopis juliflora,* 218) is common on dune areas and in many low flats and slopes along the major drainways. Cacti are seldom as common in the Chihuahuan Desert as in some of the near-by Sonoran deserts of Arizona.

The sotol (*Dasylirion* of various species, 164), above mentioned, is one of the most conspicuous of the yucca-like shrubs growing on the dry rocky mesas and hillsides of the northern parts of the Chihuahuan Desert, hence this area is sometimes called a sotol desert. The plant is rather easy to identify. From the short, spongy basal trunk spring forth hundreds of long, ribbon-like, spiny-margined leaves, each about a yard long and a half-inch wide. The small whitish flowers are borne, not in an open panicle as in the yuccas, but in a dense elongate narrow "plume" atop the flower stalk, which is often 7 to 8 feet tall and emerges from the center of the dense cluster of leaves. In times of severe drought the numerous sotol leaves are chopped off and the pulpy cores or so-called "heads," sweet with sugar, are split open and used for cattle food. Prehistoric as well as modern natives used the tough fibers of the leaves for making rope, baskets, mats, and thatch. They even ate the baked or boiled spongy pulp of the trunk. From the fermented sap the Mexicans still make a potent beverage known as sotol.

Speaking of the plant cover of the southern or Mexican portion of the Chihuahuan Desert, especially those parts which lie in the states of Chihuahua and Coahuila, Dr. Ivan Johnston of the Arnold Arboretum, who spent much time in the area, says: "Two very characteristic plants are the *candelilla* (*Euphorbia antisyphilitica,* 255), source of a high-melting-point wax, and the *guayule*

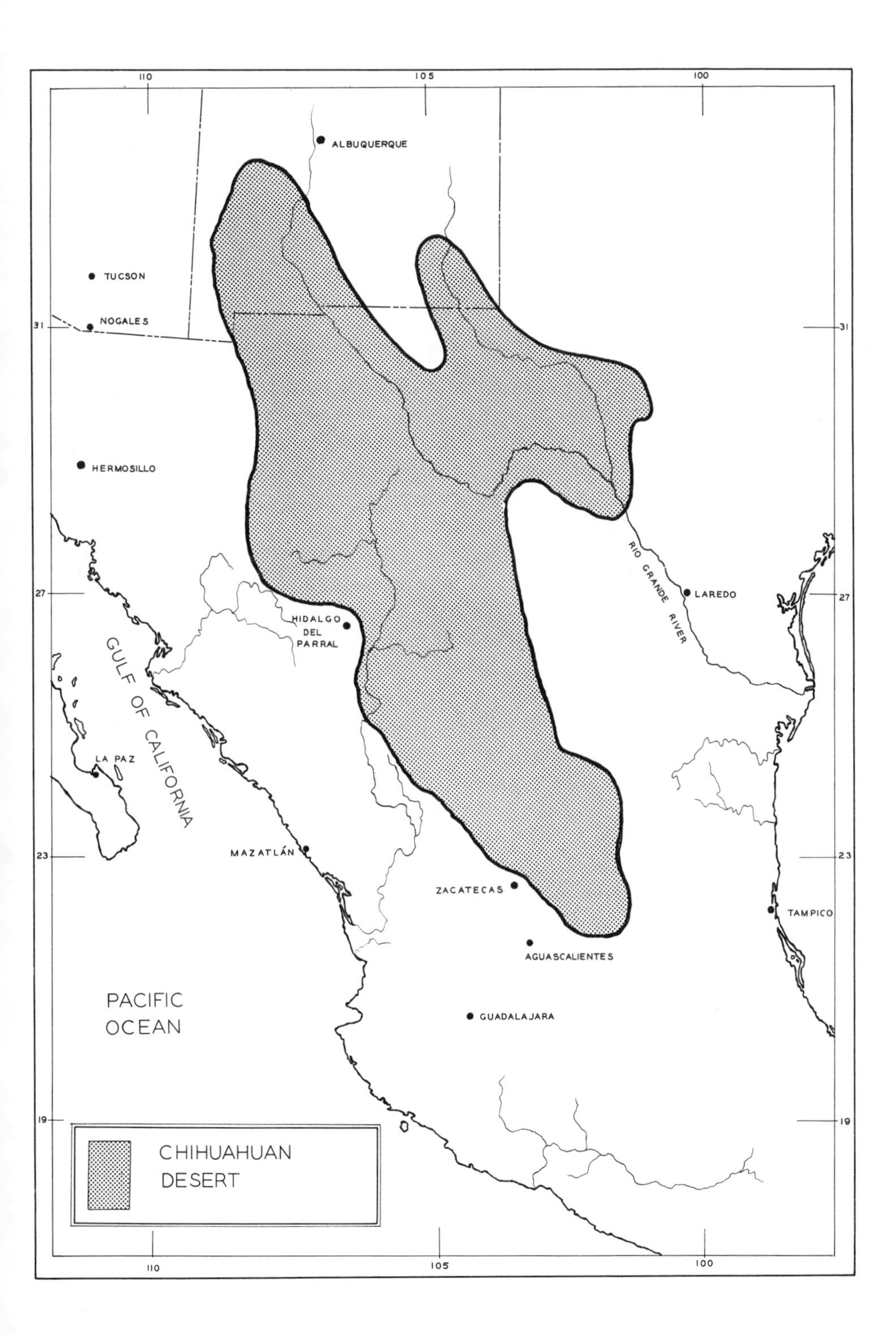
110
105
100
ALBUQUERQUE
TUCSON
NOGALES
31
HERMOSILLO
27
HIDALGO
DEL
PARRAL
RIO GRANDE RIVER
LAREDO
GULF OF CALIFORNIA
LA PAZ
MAZATLÁN
23
ZACATECAS
TAMPICO
AGUASCALIENTES
PACIFIC
OCEAN
GUADALAJARA
19
CHIHUAHUAN
DESERT

(*Parthenium argentatum*, 271), a native source of rubber. Both plants frequent limestones. The range of the guayule is practically confined to the Chihuahuan Desert. The candelilla along with lechuguilla are characteristic plants on rocky hillsides and the pediments forming the valley slopes. Over much of the region where the upper slopes in the desert are composed of limestones lies a yucca belt in which the stately *Yucca carnerosana* dominates, with its handsome and conspicuous pure-white, ball-shaped flower clusters. The guayule reaches its best development in this so-called 'palma' belt. In many areas this big yucca is the source of an excellent fiber called ixtle, and as a result the plants are grossly mutilated. The very large branching yucca, resembling the Joshua tree of California, is *Yucca filifera* (or sometimes called *Y. australis*), but differs in having the flower cluster hanging directly downward. It frequents some of the lower valleys. The Mexicans call most of the yuccas 'palma,' or the smaller ones 'palmilla' or 'palmito.' Also should be mentioned the orange-red, rather conspicuous and remarkable orchid, *Spiranthes cinnabarina*, which grows on the hot sunny limestones along with such other extreme xerophytes as lechuguilla and candelilla. On the valley floors the large patches of the foxtail-like grass, *Scleropogon brevifolius*, the 'colo del zorro' of the Mexicans, are conspicuous features. All through the Chihuahuan Desert the limestones contain various amounts of gypsum. There are many plants which are confined almost wholly to gypsum soils or which occur most frequently on that substratum. Another interesting plant of the limestone slopes is *Selaginella lepidophylla* ('Flor del pena'), which grows with lechuguilla and candelilla. During dry weather it is rolled up into a tight ball but after rain unfolds to form a nice green rosette."*

The name Tularosa Basin or Tularosa Desert is given to the northernmost parts of the Chihuahuan Desert lying to the west of Alamogordo, New Mexico, between the Rio Grande and Pecos rivers. It is described here at some length because it contains many of the desert's most unusual and interesting features.

Geologists believe the Tularosa Basin was formed by the sinking of a narrow block of a former high plateau. This explanation seems reasonable since thick beds of gypsum found beneath this desert basin are matched by remnants of similar or identical beds found high above the basin floor in parts of the old plateau which evidently did not settle and which now constitute the mountains flanking the basin to the east and west.

* Quoted from a letter from Dr. Johnston to the author.

Among the Tularosa Desert's unusual features are the ancient beaches of the extensive prehistoric Lake Otero, which once had an outlet to the Rio Grande; also attracting attention are the flat-bottomed arroyos (known as lost rivers), extensive salt marshes, gypsum beds, alkali-encrusted clay pans, widespread lava flows, and dazzling white sands forming extensive dunes, once called the Gypsum Hills. This great area of dunes lies on the east side of the San Andreas Mountains, at whose base is a particularly large, barren alkali flat. At the southwest end of the dunes lies ephemeral Lake Lucero, which long has served both as a catchment basin for gypsum-bearing waters draining from the near-by mountains and as a great evaporation pan, especially during summer. Prevailing southwest winds pick up the white gypsum particles left by the evaporating waters and deposit them in the dune area, augmenting the accumulations of past centuries.

The almost sugar-white sand is composed of nearly pure gypsum (*calcium sulphate*), the mineral from which plaster of Paris is made. In its massive, compact, translucent form, calcium sulphate is known as alabaster. To the north of the white gypsum deposits is another dune area almost as large but made up of gray quartz sand of granitic origin.

The sands of this drifting gypsum wonderland are often disposed in mamelons (rounded hillocks) and ridges from 30 to 50 feet high. They cover an area of nearly 400 square miles, well over half of which are included in the White Sands National Monument, lying about 15 miles southwest of Alamogordo.

The wind-rippled sands, constantly shifting to the eastward, are most appealing on the warm moonlight nights of midsummer. One may then stroll on the dunes for hours, enjoying their marvelous cleanliness, the soft flowing lines, and the snowlike appearance. The southwest breezes steadily sweep over the surface, ever forming new ripple patterns. By day the dunes are of special delight to artists and photographers, particularly when there is a low sun to bring their patterns into sharp relief.

The most noticeable large plant of the almost snow-white dunes is the vigorously growing, aromatic three-leafed sumac (*Rhus trilobata*), a shrub that is widespread in desert lands. This sumac is often known as squaw bush because many Indians formerly utilized its withe-like branches in making baskets. Here it usually occurs, not in thickets, but as scattered single hemispherical bushes, growing either at the crest of sand hummocks or riding on the tops of strange "sand columns." These columns are really pillars of granular gypsum, some of them 15 feet high, held in form, in spite of the work of constant winds, by the shrub's numerous roots which freely branch and penetrate through them. Each such column-crowning squaw bush was once growing on the top of a dune. The dune, still active, gradually moved on, leaving stranded the root-bound gypsum pedestal.

Another common dune plant is the narrow-leafed yucca (*Yucca elata*, 161), known as palmilla (Spanish, "small palm") or soap weed in allusion to the saponaceous qualities of its crushed roots. Here it grows with most of its stalk buried beneath the sands. Out on the adjacent open desert it is a handsome and conspicuous small "tree," often attaining a height of 10 to 12 feet, not counting the numerous, much higher extending flower-stalks, which in spring carry tall terminal panicles of handsome white flowers.

A large green, broom-like desert tea (*Ephedra*, 208, 210), one of the most peculiar of the conifers, is often found on the surface of the dunes as well as in the low bottoms between. It, too, may have only a small portion of its upper branches showing above the surface of the constantly shifting sands. Other woody plants trying to grow upward fast enough to keep from being smothered under the moving sands are several kinds of salt bushes (*Atriplex*) with almost silvery white foliage, a cottonwood tree (*Populus fremonti*), a shrubby pennyroyal (*Poliomintha incana*, 329), and two kinds of rabbit brush (*Chrysothamnus*). Between the sand hillocks grows a beautiful loose-panicled grass, alkali sacaton (*Sporobolus aroides*). Individual plants of at least seven species, which usually grow to be only a few feet high under ordinary conditions, here grow thirty or more feet upward through the dunes, developing at the same time many adventitious or specialized side roots along their long extended stems.

The most abundant plant cover is found in the more stable soil in the irregular flat bottoms between the dunes. Here grow almost turf-like stands of grama grass (*Bouteloua*) and low bushes of squaw tea (*Ephedra*, 208, 210) as well as numerous small annuals and perennials. According to Dr. Fred W. Emerson, who made special studies of the plant life of the White Sands, "at least 55 species are limited to the flats because they cannot elongate their stems when covered" with the encroaching sands. He states that "there is no evidence of any reproduction of any plant except in the flats. The motion of the sand and lack of available water in the superficial layer of the active dunes conspire to prevent the growth of seedlings. In the flats, however, seedlings mature, and

New Mexico State Tourist Bureau

The yucca, New Mexico's state flower, even grows in dunes of the White Sands. This expanse of almost pure gypsum covers an area of 270 square miles in southern New Mexico. A portion of the White Sands has been set aside as a National Monument, with headquarters southwest of Alamogordo.

New Mexico State Tourist Bureau

White Sands National Monument, near Alamogordo, New Mexico. The eastern edge of the Sands. No vegetation is found except along the edge. The formation is actually composed of granulated gypsum and contains no sand whatsoever.

New Mexico State Tourist Bureau

White Sands National Monument in southern New Mexico appears to be a vast expanse of snow from the heights of the Alamo fire station in the Lincoln National Forest. The gypsum mass has been piled into great dunes from 10 to 60 feet high by the prevailing southwesterly wind.

fruiting plants are very numerous wherever vegetation has penetrated. Because the plants are rooted in the flats it is possible to determine accurately the total height of plants that are partly covered by dunes."

In April and early May and again after summer rains, an abundant representation of beautiful and often rare flowering annuals and grasses is to be found on the dunes. Because of the unusual physical and chemical properties of the almost nitrogen-free soil to which they have adapted themselves, they show many unique characters. Among these is the ability to absorb water when their roots are surrounded by almost saturated solutions of gypsum.

On the dunes there is a unique kind of plant succession. First appear the so-called "pioneer plants." These are supplied the necessary nitrogen by peculiar primitive nitrogen-fixing plants (mycorhizas) growing on their roots. The yucca (*Yucca elata*, 161) is such a pioneer plant. After the death and decay of these "pioneer plants," other plants which succeed are believed to derive their nitrogen supplies from this decomposed organic material.

The soil of the White Sands consists of 94% granular gypsum, 3% table salt, and 1% silicates. One wonders how any plant can gain a foothold in such a soil, let alone flourish there. Of course only those which have a root system highly adapted to a saline substratum can survive. Of water there is always enough, because gypsum, which is soluble in cold water, long holds the greater part of the moisture that falls upon it. Dr. T. D. MacDougal once referred to these dunes as mere "moist hillocks of granular structure." Only the very surface layers are dried out by the heat of the sun, and then only to a depth of a few inches. This thin dried upper layer is constantly being lifted off by the strong prevailing southwest winds, leaving the newly exposed layers to be dried out in turn.

Many white-footed mice and other kinds of desert rodents subsist on the seeds of the numerous plants. They, in turn, are the prey of the large-eared, cat-size kit fox (*Vulpes macrotis*, 125).

Just a few miles north of White Sands is another desert area of singular beauty and of great interest to the biologist. It consists of a great black lava flow of comparatively recent origin, covering an area of some 120 square miles. This is called the Tularosa Malpais (Spanish, "badland"). The source of this flow is a low cone at the north end of the lava beds, a few miles northwest of the town of Carrizozo. The congealed lava varies in thickness between 10 and 50 feet; the flow itself is seldom more than 5 miles wide. At the southern edge is a copious spring of alkaline water, Malpais Spring, in which lives a peculiar cyprinodont minnow, a relict species which is thought to have lived here since the prehistoric time when the spring was larger and had a direct connection with the Rio Grande.

The northern portion of the Tularosa Malpais is 1,200 feet higher in elevation than the southern end. There is, as one might expect, a considerable difference in the plant cover as you traverse the 45-mile length of the lava flow. Here you go northward from typical low mesquite–salt bush–creosote desert at the southern lower end into a regular grassland association on the edge of a juniper-sotol belt of vegetation.

This region is of special interest to the biologist because here we have the black lavas contiguous with the almost pure white gypsum sands, and identical species of animals on both areas. Scientists have long queried whether the color of the soil might in any way influence the color of an animal living on it. As a result of detailed studies made on the Malpais, it was found that five of the commonest small mammals living in this area, all rodents, are definitely much darker on the Malpais than are identical species living on the adjacent white sands. These darker furred rodents are the rock pocket mouse, rock squirrel, juniper mouse, white-throated woodrat, and Mexican woodrat. Another species, the cactus mouse, was found to be dark-colored in some cases and light-hued in others, indicating probably that it is among the most recent rodents invading the lava beds.

Wide-ranging species, such as the cottontail, jack rabbit, and coyote (which are not actually resident on the Malpais but merely forage over it in quest of food), were, as would be expected, all light-colored.

Protective coloration has long been a conjectural topic among biologists and the final word has not yet been said about it. It is generally believed, however, that when a species invades a dark-soiled region, any of its offspring that are a bit darker in coloration will stand a much greater chance of escaping detection by predatory enemies and thereby have a much better chance of surviving than the paler forms. It is assumed that after a long period of time, through much inbreeding of the surviving dark specimens, a race that is almost black ultimately will be derived. The reverse would be true if a dark-pelaged animal invaded the whiter desert soils. In the White Sands lives an almost pure white pocket mouse, surely the ultimate in protective coloration no matter how evolved! In the adjacent red hills, pocket mice with red pelage are known.

Those creatures which are fully adapted to living on the light desert soils but which still retain sharply contrasting colors, like the darkling beetles (*Eleodes*, 11), or the velvet "ants" (*Mutilla*, 24), may have kept their seemingly incongruous color as a warning device or as an advertisement of the fact that they are either bad-tasting or poisonous. In the case of the black beetles their color is possibly a protection against the sun's ultraviolet rays.

To the west of the White Sands rise the serrated San Andreas Mountains, composed of banded limestones. Just to the west of these is the famous bolson bearing the suggestive and doleful name of Valle Jornada del Muerto (Valley of the Journey of Death), a name bestowed upon it because of the number of early Spanish refugees who perished there of thirst in 1690 while in flight down its waterless length toward Old Mexico. Across the valley's 90 miles of sere desert and rough lava beds passed the first group of 400 colonists as they traveled in 1598, under the leadership of Juan de Oñate, to establish the first permanent colony of Caucasians in New Mexico.

A well-known undrained basin of the Chihuahuan Desert, lying in Texas north of the Rio Grande between the Guadelupe and Hueco mountains, is the Diablo Bolson. It is really a mountain-bordered plateau with an elevation of 3,500 to 5,000 feet and a rainfall of 7 to 8 inches. The large salt flats at the foot of the Guadalupe Mountains were for more than two centuries the source of salt for people dwelling over a wide area in the upper Rio Grande Valley and northern Mexico; for all purposes it was considered as common property. In 1887, when an attempt was made to file claims on the salt beds and convert them to private ownership and use, bitter disputes arose which led to the famous "Salt War" in which a number of citizens were killed.

The Mexican portion of the Chihuahuan Desert is mostly a vast elevated intermountain plateau, much of it built by the outpouring of volcanoes and by the deposition of calcareous debris brought down from neighboring limestone mountains by intermittent streams. Its broad surface is broken frequently and irregularly by low, often barren mountains and their broken and much eroded pediments, with increases in elevation toward the south; much of it lies at an elevation of from 3,000 to 5,500 feet. The basins or valleys between the mountain ranges impress one continually by their great broadness. Distant mountains seem very far away. The lofty Sierra Madre Oriental, composed largely of Cretaceous limestone, bounds it on the east; to the west lie the higher, rugged Sierra Madre Occidental, produced by massive block faulting, volcanic extrusion, and later severe erosion; the latter range, together with related mountains closely situated to the east, form a part of the Continental Divide. These high mountains, which guard the desert's confines on the east and west, shut off rains from both directions and thus account for the extreme aridity of this vast inland area. The fitful and rather badly distributed rainfall averages between 2½ and 10 inches annually. The winters are on the whole dry. The end of June is the hottest part of the year, especially on the desert plains. July, August, and September span the season of greatest precipitation; the summer rains blow in from the Gulf of Mexico when the tropical or easterly wind system has greatest influence, and with them comes the summer-autumn wild flower season. The grasslands now change into beautiful green meadows, later to become handsome fawn-colored swards.

The only river of any consequence running through this great intermountain plateau, which measures 200 to 300 miles across from east to west and fully 800 miles long, is the cottonwood-bordered Conchos. It arises in the western sierra and flows northeastward into the Rio Grande.

In the northern part of the State of Chihuahua are several smaller streams which have their sources in the oak-grassland regions of the elevated plains to the west of the creosote bush desert. They once flowed into the Rio Grande, but now drain into three large lake beds or "pans": Laguna de Gusman, fed by the Río Casas Grandes; Laguna Santa María, receiving water from a river by that name; and Laguna de Patos ("Lake of Ducks"), near Villa Ahumada and getting water from the Río Carmen. Fed by heavy summer rains and flowing at a steep gradient, these small streams carry large loads of sand and silt on through to their terminal sinks. When, as sometimes happens, the water entirely disappears from these lake beds, the strong winds pick up the silts and sands, often carrying them for distances of 50 to 60 miles. Some of these wind-blown deposits form the extensive chain of dunes found near Samalayuca (elevation 4,360 feet), about 30 miles south of the International Border. Some of these dunes are crescent-shaped, high, and migratory; these are called "barchans" by the geographers. Others are more or less stable and form a broad complex of hillocks and high barren ridges which rise some 200 to 300 feet above the desert plain. Unlike the gypsum dunes of the White Sands area in New Mexico, the material composing these Mexican dunes is 94% silicon. They are beautiful and imposing sights at any time, the lower dune areas especially so when after favorably spaced and copious summer rains they are carpeted with a luxuriant growth of wild flowers.

Growing on the most stabilized of these sand hummocks can be found open stands of honey mesquite (*Prosopis juliflora*, 218), squaw bush (*Rhus trilobata*), fine-leafed wormwood (*Artemisia filifolia*, 184), snowy bush-mint (*Poliomintha incana*, 329), Torrey squaw tea (*Ephedra torreyana*), fourwinged salt bush (*Atriplex canescens*), palmilla (*Yucca elata*, 161), the magenta-flowered, broom-like dalea (*Dalea scoparia*), and the blue-flowered Thurber penstemon (*Penstemon thurberi*, 338), all of them shrubs found also in the deserts of Arizona and New Mexico immediately to the northwest. Most of the more massive dunes are almost devoid of plants except after summer rains.

Far to the south and east of the Río Conchos are other large undrained basins with dry lake-beds of alkaline salts and limestone clays in their lowest parts. The most famous—and long the most dreaded by early travelers—is the enormous, broad Bolsón de Mapimí. This bolson once contained in past geologic time a large body of water. Much of it is still a wilderness land, known intimately only by miners and cattlemen, the monotonous domain of creosote bushes and other strictly desert shrubs. Salt bushes (such as *Atriplex canescens*), inkweed, mesquite, and other alkali-tolerant plants grow sparsely to thickly around the edge of the seemingly endless sun-blistered flats and playa deposits of fine-textured calcareous clays. Groups of dust devils, or tornillos, rise frequently all during the heated days, carrying the white and brown alkaline dusts majestically upward and across the land in giant spiral columns. Miles and miles away one may see through the quivering air of the summer sun the heat-distorted outlines of the bordering mountains; on their adjacent bajadas and pediments grow the honey-bean mesquite, various forms of acacia and agave, and the massive, wide-leafed yucca (*Yucca carnerosana*). But the heat of this region is not comparable with that of the far western deserts; the area is, during most of the year, much cooler because of its elevation.

At the lower end of the Bolsón de Mapimí lies Laguna de Mayrán, a broad depression which has for many ages received the waters and loads of silt of the Río Nazas, flowing eastward and southeast from the highlands of the Continental Divide. Other streams temporarily flowing westward into the Laguna have added their share of detrital materials so that now the southern bolson bottom is a vast, rich, and deep-soiled flatland area which has in recent years encouraged the practice of agriculture on a large scale. The city of Torreón lies at the western side of this cultivated area (now known as the Laguna district). It has become the metropolis of one of the most populous areas of Mexico.

The plains and basins of the Chihuahuan Desert in the State of Coahuila are punctuated or bounded by a considerable number of low shrub-covered or rather high oak-and-pine–covered mountains. The broad basins and valleys of the southern, western, and northern three-fourths of the state are strictly desert, where the most common hardy plants are the creosote bush and such associated shrubs as the varnish bush (*Flourensia cernua,* 265), varnish-leaf acacia, ocotillo, and mesquite. Here one also finds, especially on the higher slopes, the broad-leafed Torrey yucca (*Yucca torreyi,* 162), all-spine (*Koeberlinia spinosa,* 147), and the thorny condalia (*Condalia lycioides,* 144). The soils where these grow is usually rather shallow or stony, and may even be covered by desert pavement. In places where the soil is deep and well drained the yellow-flowered flourensia becomes abundant, perhaps actually displacing the creosote bush. Especially well-defined thickets of varnish-leaf acacia (*Acacia vernicosa,* 192), small-leafed sumac (*Rhus microphylla*), and the tree-like *Yucca australis* occur in places. This vegetation develops on less well-drained local depressions, in contrast to the concentrations of flourensia. Such shrub thickets are locally called mogotes, a Spanish word meaning islands, hillocks, or mounds.

Northeastward from Durango to Torreón is a broad arid valley with elevations between 6,000 and 7,000 feet. Cacti, especially barrel cacti, large-padded opuntias, and cane opuntias, called cholla by the Mexicans, are common. About 100 miles north of Durango a steep escarpment drops suddenly down into a broad desert valley filled with vast assemblages of creosote bushes and small agaves. Eastward the country becomes more and more arid until Torreón is reached. East of the flats of Laguna de Mayrán the land rises very gradually for a hundred miles or more as it passes through a comparatively barren creosote-bush desert with an average rainfall of but 3 inches. Mountains running in an east and west direction and of considerable heights can be seen south of the highway; to the east, the Sierra Madre Oriental gradually comes into view, with the city of Saltillo (elevation, 6,212 feet) near its west base. North of Saltillo the mountains become much lower. East of Monclova, the Río Salada, which rises in central Coahuila, has eroded a wide pass through the mountains, forming an area of low elevation cutting into the plateau. After the enticingly beautiful Cañon de San Lazaro the vegetation becomes still more dense; the abundance of cacti, yuccas, acacias, and herbaceous flowering shrubs shows the effects of greater precipition. This is another Mexican region where the summer rains are often torrential, the product of storms blowing in eastward from the Gulf of Mexico.

North and east of the city of Chihuahua is a large area of outwash plains covered in many places with almost pure stands of the ever-frequent creosote bush. Sometimes associated with it are honey mesquite and the low-statured tar bush (*Flourensia cernua,* 265), best developed on deep soils. On the lower bajadas, in addition to these three dominant shrubs, there may be found such woody plants as crucifixion thorn (*Holacantha emoryi,* 336), cat's-claw (*Acacia greggii,* 221), two species of condalia (*C. lycioides* and *C. spathulata,* 148), and the small shrubby varnish-leaf acacia (*Acacia vernicosa,* 192), found also in western Texas and southern Arizona.

Northeastward and higher on the slopes of the mountains are such dominant desert-shrub plants as the stately yucca (*Yucca carnerosana*), the sotol (*Dasylirion wheeleri,* 164), the lechuguilla (*Agave lechuguilla,* 197), the handsome shrubby Gregg coldenia (*Coldenia greggii*), with its headlike clusters of bell-shaped flowers that range in color from pink to magenta, and the flexible-stemmed leather plant (*Jatropha spathulata,* 146), called by the people of Durango and Chihuahua "Sangre de Drago" ("Blood of the Dragon") in allusion to the red dye made from its bark. In places a slender-stemmed form of the ocotillo (*Fouqueria splendens*) forms wide dense areas of "elfin forests." Tussocks of coarse "tobosa" grass (*Hilaria mutica,* 137) and other grasses quite often cover the ground as an open sod in the intervening spaces. The only conspicuous cacti here are both species of opuntia. One is the purple-flowered cane cactus (*Opuntia imbricata*), of almost treelike proportions (often 5 to 6 feet tall) with yellow fruits; the other is a ground-hugging, flat-jointed, yellow-flowered prickly pear (*Opuntia macrocentra*).

To the southwest of Chihuahua city and west of the cottonwood-bordered Río Conchos and its branch, the Río Florida, is another rolling limestone plain rising toward the west. It has an elevation varying from 3,800 to 4,500 feet and extends south almost to the picturesquely situated mining town of Parral. Much of the soil is extremely shallow and in many places the white or gray limestone is exposed. In the drainways of little depth and in the larger depressions are found dark gray clays where dense thickets of the varnish-leaf acacia (*Acacia vernicosa,* 192) grow. Here and there are shallow depressions on whose broad flat bottoms water sometimes may stand. These areas are often covered with the coarse nutritious grass (*Hilaria mutica,* 137) called "tobosa" by the cattlemen. The two most noticeable cacti of these limestone plains are flat-jointed prickly pears. One, *Opuntia engelmannii,* is a fairly large, wide-spreading, shrubby species forming bushes 5 to 6 feet high and often up to 5 feet across. It has pale-green joints, short spines, and large yellow flowers fully 3 inches in diameter. The other cactus, the red or purple-tinged prickly pear (*Opuntia macrocentra*), common also in west Texas, southern New Mexico, southern Arizona, and northern Sonora, is a smaller shrub with dark-red or purplish rounded joints, covered sparingly along the upper margins with very long, slender, deep red to nearly black spines. The flowers, also yellow but with red bases, are somewhat smaller, only 2 inches in diameter.

Mammals common to the Mexican part of the Chihuahuan Desert include the Mexican prairie dog (*Cynomys mexicanus*), huge-eared, white-hipped, Gaillard antelope jack rabbit (*Lepus gaillardi*) found also in southern New Mexico, the kit fox (*Vulpes macrotis zinseri*), the banner-tailed kangaroo rat (*Dipodomys spectabilis*) with prominent white-tipped tail, and in the higher arid mountains the Mexican bighorn (*Ovis canadensis mexicana*).

One of the best means to get acquainted with a representative bit of the United States portion of the Chihuahuan Desert is to visit the Big Bend National Park in southern Texas, where the Rio Grande makes a large U-shaped bend in its meandering course. One can go also to Carlsbad Caverns National Park, in the vicinity of which typical Chihuahuan desert vegetation and animal life may be seen in usual abundance. To the west is the White Sands National Monument of the Tularosa Basin, previously described. All of these nature sanctuaries are open to visitors throughout the year and good roads make them readily accessible. A walk over one of the several nature trails where the outstanding local plants are tagged with names is a rewarding experience.

Much of the Mexican portion of the Chihuahuan Desert is at

present inaccessible. At this time only three paved roads lead through it: one is the well-surfaced Christopher Columbus Highway; the second is the highway from Durango northeast through Torreón, Saltillo, and north to Eagle Pass, Texas; the third is the highway from Lagos de Moreno southeastward through the states of Guanajuato, Querétaro, and Hildalgo.

In the last few years considerable plains portions of the erstwhile desert have been supplied with irrigation water and are now used for the growing of maize and cotton. Not only are the cities along the route such as Chihuahua, Parral, Durango, and Zacatecas most interesting but also the small towns—Nombres de Díos, Sombrerete, Vincente Guereró, and Lagos de Moreno, with beautiful old churches and colorful markets where native wares are on display. The Mexicans are everywhere a kindly, helpful people; traveling among them is always a delightful experience. It is an old saying that "once the dust of Mexico has settled upon your heart you can find peace in no other place."

4

THE *Sonoran* DESERT AND ITS SUBDIVISIONS

The Sonoran Desert is most characteristically developed in the western half of the Mexican state of Sonora, hence the name. In somewhat altered form, it extends into southern Arizona, southeastern California, and much of the upper three-fourths of the peninsula of Baja California. All of these areas immediately surround the Gulf of California (see maps, pages 52, 62, 72, 84, 104).

This desert consists of broad, sandy, and often rocky arid plains more or less isolated by numerous barren and often detached mountains. The eroded materials carried down from the mountains, mostly by the waters of violent storms, form broad, outwash detrital fans or bajadas which gradually slope down to the centers of the wide plains. Large washes, and sometimes small rivers such as the Yaquí, carry the runoff waters to the Gulf of California or into the bottoms of sinks, where they form ephemeral lakes. The large Colorado River cuts directly across parts of this desert and receives drainage from them. Because of the abundance of small trees, in contrast to the numerous low shrubs of the Great Basin Desert further north, the Sonoran Desert is often spoken of as an arboreal, or tree, desert. Sizable trees and tree-like cacti are particularly well developed here. There is a great display of succulents, large and small, and great numbers of evergreen and deciduous shrubs. The rainfall averages from a mere trace in some of the western parts, receiving only winter rains, to 13 inches on the eastern edge, where both summer and winter precipitation occur.

For convenience in describing this vast arid region of nearly 120,000 square miles, with its varied vegetational and animal features and its strange scenic contrasts, the Sonoran Desert is treated under six major divisions: the Plains and Foothills of Sonora; the Arizona Upland or Sahuaro Desert; the Yuman Desert; the Colorado Desert of California and areas surrounding the upper part of the Gulf of California; the Vizcaíno-Magdalena Desert of Baja California; and the Gulf Coast Desert.

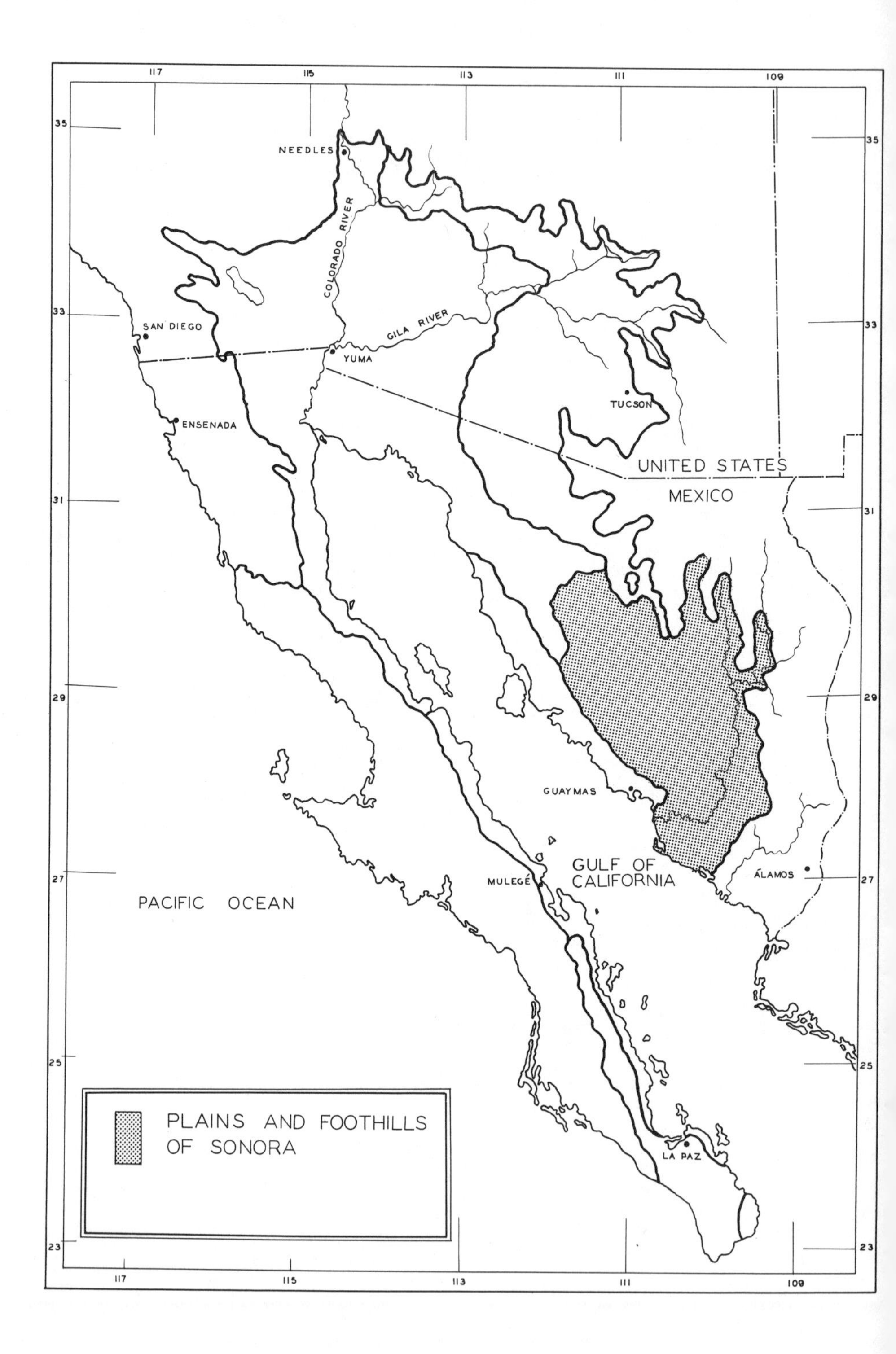

117
115
113
111
109
35
33
31
29
27
25
23
NEEDLES
COLORADO RIVER
GILA RIVER
SAN DIEGO
YUMA
TUCSON
ENSENADA
UNITED STATES
MEXICO
GUAYMAS
GULF OF CALIFORNIA
MULEGÉ
ÁLAMOS
PACIFIC OCEAN
LA PAZ
PLAINS AND FOOTHILLS OF SONORA

5

THE *Desert Plains* AND *Foothills* OF *Sonora*

A journey southward into Mexico over the highway between Nogales and Guaymas takes one across the undulating desert plains of Sonora and even into a bit of the Gulf Coast Desert. It is a most interesting trip because of the variety and abundance of new and strange plants, especially trees and shrubs, which gradually make their appearance. Much of the desert along the highway, as far south as Magdalena, lies within a belt running more or less north and south in which the yearly rainfall measures between 10 and 15 inches. This is an amount about equal to that of the Tucson district in Arizona. South of this the rainfall decreases rapidly but still is sufficient to encourage the growth of shrubs and low-branching trees, which in many places form extensive open scrub- and small-tree forests. Palo verdes, ironwoods, and honey-bean mesquites here find conditions most favorable to healthy growth and attain sizes not often found in other portions of the great Sonoran Desert. Creosote bush and burro bush are no longer the dominant shrubs, nor do they usually occur in such large pure stands as they do in the deserts of the lower Colorado Valley. Here in Sonora they are generally found only in scattered colonies, alternating with the attractive open forests of low trees. There are many interspersed groups of the organ pipe cactus, senita cactus, and ocotillo. The sahuaro is found less and less frequently and south of Hermosillo is quite rare indeed, its place there being taken by the large, bulky-stemmed *Pachycereus pecten-aboriginum.* Wherever there are open spaces, the brittle bush (*Encelia farinosa,* 267) and, in places, grasses take over to form an open ground-cover.

Upon reaching the central third of the Sonoran plains, near Hermosillo, one readily becomes aware of a new and strong enrichment in the composition of the flora. Almost suddenly there appear, growing in dense thickets, especially on the higher slopes, a number of trees and shrubs unique in form and color. Among these are the peculiar tree ocotillo (*Fouquieria macdougalii*) with firecracker-shaped, deep-red flowers, and the delicate, vividly green-barked Sonoran palo verde (*Cercidium sonorae*). There now begins to appear the strange white-barked tree morning-glory (*Ipomea arborescens*), which is entirely leafless in the dry season. The showy flowers appear after the rainy season begins and while the tree is still leafless. The odd, deep-green jito tree (*Forchammeria watsoni,* 142) is found among the mesquites, acacias, and cacti on the plains; and the delicately branched, white-barked palo blanco (*Acacia willardiana,* 220) grows in scattered colonies on the rocky hillsides. If it be winter or spring, the general color tone of the landscape is likely to be a combination of gray or yellow-green and soft pastel brown. But if there have been particularly generous winter rains, then some of the trees and shrubs may put on a temporary show of new green leaves and even of flowers during late March and April. The all-important summer rains generally begin about the end of June. (The Mexicans are more specific and say it is on June 24, St. John's Day.) Soon thereafter, grasses and many varieties of annual wild flowers carpet the ground, and many of the herbaceous perennials and trees, which have been up to that time inactive, abruptly break into full leaf and flower. The whole appearance of the desert now has changed within a period of days, and it has become a place of richest verdure, with brilliant patches of color on every hand. In the thickets and along ordinary dry streamways, numerous flowering vines now magically appear and climb to the very tops of the shrubs and trees, adding their share to the riot of floral color.

Certain to attract attention is that remarkable and peculiar relative of the cucumber, ibervillea (*Maxicmowiczia sonorae,* 240), with its huge, half-exposed or subterranean, thick water-storing stem and globose root. It is generally discovered beneath shrubs or trees. So much liquid is stored in this unique plant reservoir, during times of plentiful rainfall, that its vines grow up, flowers bloom, and small plum-sized fruits mature during several successive seasons of severest drought. The vines grow up through the trees under which they grow. The salver-form flowers are yellow, the fruits a small round, melon-like red berry. Other species of ibervillea are found in the deserts of New Mexico, Texas, and Chihuahua.

Branching off from the paved highway are many little-used side-roads leading off into virgin stands of "dwarf forest" trees. The wise traveler, exercising the usual caution when going over unimproved roads, will do well to seize some of these opportunities to explore the hinterland. Even a drive of half a mile from the paved highway may bring rare chances to see much that is new, exciting, and mentally stimulating. At lunchtime let him picnic in the open groves of trees, and if possible he should camp for the night where he can prepare his dinner over a fire made from the unfamiliar dead wood lying about, noting, as he cooks, the new and appealing odors of the smoke as well as the varied colors of the ever-restless flames. When he awakes in the early morning he will hear many new bird songs, perhaps even glimpse some of the strange songsters. Later as he walks abroad he will be amazed at how many new plants there are to be seen, plants he would never notice while traveling at high speed down the major highway.

Four principal rivers drain the slopes of Sonora. All flow through the deserts in a southwesterly direction. It is only during the occasional heavy rains of the wet season that their waters reach the sea. Extreme southern Sonora is drained by the Río Mayo. The Río Yaquí and its tributaries drain most of the eastern and southern portions. Most of north-central Sonora is drained by the Río de Sonora and its tributary the Río de San Miguel. The Río de Magdalena carries away runoff water from a large area of the arid northwestern portion.

It is along these rivers that most of the larger rural settlements occur. There are poor roads which lead along the streamways from one settlement or village to another, but these make it possible for the traveler to penetrate far into primitive areas where he will be able to observe the native peoples, their quaint villages and small farms. Here is Mexico at its best.

To the east of the Sonoran plains is a rugged but not mountainous surface designated as the Sonoran foothills, really foothills of the high Continental Divide. Here too there is a long period of deficient rainfall followed by three months of rain in summer. In many ways this region is much like the Arizona Upland Desert near Tucson and has about the same amount of rain. Here we find many shrubby cacti, often in quite dense thickets. Especially common on the rolling hills are many low-statured trees, mostly small-leaved leguminous species such as many kinds of acacias—also mesquites and palo verdes. The sahuaro and organ pipe cacti continue to flourish here, as do also two species of ocotillo and the ubiquitous brittle bush (*Encelia*). In canyons may be seen the Sonora fan palm (*Washingtonia sonorae*). Its distinguishing character is found in the leaf stem, which is obtuse at its junction with the leaf blade and not pointed and prolonged into the blade as in the California fan palm of the Colorado Desert.

The Sonora foothill desert is, as a whole, rather poorly supplied with roads and most of these are unimproved and generally in a very unsatisfactory condition, especially after heavy rains. Consequently many of its most colorful and interesting parts, such as are found in the black lava beds and varicolored volcanic hills, are virtually inaccessible except by muleback, horse and wagon, power truck, or jeep. Wood roads and trails known only to the native peoples lead far back into the mountains covered with oaks and pines.

The hills of the southern part of this upland desert are heavily clothed with sizable trees and tall grotesque shrubs, some of them species which have migrated northward from the more humid tropical thorn forest. Trees of considerable stature are most prevalent, as might be expected, along the margins of streams.

In connection with trees of special interest, there needs to be mentioned the remarkably beautiful bignoniaceous amapa (*Tabebuia palmeri*), which sometimes grows to a height of 60 feet and has masses of magenta- to peach-colored flowers occurring in small, dense, showy head-like clusters. Like those of the morning-glory tree, its blossoms may occur while the tree is still leafless. It occurs in the rolling hills to north of Cuidad Obregón and southward through Alamos and into Sinaloa.

Always attracting attention because of its yellowish, birch-looking, shedding bark and rubbery branches is the tree-like jatropha (*Jatropha cordata*) found so often in the central Sonora plains with tree ocotillo and ironwood. It may be barren of leaves during most of the winter and spring.

Found in similar situations is the shrub-like and rubbery tough-stemmed sangregrado (*Jatropha spathulata*, 146), mentioned also as a prominent plant of the Chihuahuan Desert flora.

Visitors who have traveled as far south as Guaymas on the Gulf of California may wish to go on southward across the delta of the Yaquí River to Cuidad Obregón, on to Navajoa, then go eastward to spend some days in the charming old silver-mining town of Alamos. En route they may view the rapid transition

Plains of Sonora with a rich shrub flora consisting of many low-flowering jatrophas, brittle bush (*Encelia farinosa*), acacias, and the larger organ pipe cactus (*Laimereocereus thurberi*), tree ocotillo (*Fouquieria macdougalii*), and ironwood trees (*Olneya tesota*).

William Woodin

Vast forests of the giant cactus (*Cereus giganteus*) in the vicinity of the Arizona-Sonora Museum near Tucson, Arizona. The dense trees are mostly little-leaf palo verdes (*Cercidium microphyllum*).

William Woodin

Arboreal desert, southern Arizona. Giant cacti (*Cereus giganteus*), ironwood trees (*Olneya tesota*), and many shrubby cacti and thorny bushes are widespread.

from austere desert to the thick jungle of the semitropical thorn forest, where there is an increasing complexity of composition in the xeric plant cover.

The strange heavy-thorned acacia (*Acacia cochliacantha,* 222) now becomes one of the commonest trees. The large day-flowering tree cactus (*Pachycereus pecten-aboriginum*),* with much the stature of the cardon, is widely scattered among the other sizable strange new trees which here abound. Flocks of screaming parrots may fly overhead and handsome long-tailed magpie jays sharply call from the treetops.

Among the larger birds of interesting habits that may be seen in the desert portions of Sonora are the masked bobwhite (*Colinus ridgwayi*), Gambel quail (*Lophortyx gambeli,* 73), Mexican sparrow hawk (*Falco sparverius,* 67), poorwill (*Phalaenoptilus nuttallii,* 83), white-necked raven (*Corvus cryptoleucus,* 84), Mexican raven (*Corvus mexicanus*), roadrunner (*Geococcyx californianus,* 79), Mexican ground dove (*Columbianus passerina pallescens,* 92), the caracara (*Polyborus cheriway auduboni,* 65), turkey buzzard (*Cathartes aura,* 69), and black vulture (*Coragyps atratus*).

Because of the great number of available nesting sites and the quantity of food in the form of wild seeds and insects, the number of smaller birds is many. A goodly proportion of these are species more or less common in the mesquite and grassland desert areas of southern Arizona; others are unfamiliar species which enter the Sonoran Desert plains from the tropical lowlands of the State of Sinaloa to the south. Here the tropical and desert elements merge.

A bird certain to attract the attention of the traveler in Sonora, as well as in northern Baja California, is the odd-looking Audubon caracara (*Polyborus cheriway auduboni*), locally known in southern Arizona as the Mexican buzzard because of its habit of feeding on carrion (the generic name, *Polyborus,* is a Greek-derived word meaning voracious). This bird of prey, closely allied to the falcon, occurs in considerable numbers: it will probably be first noticed while conspicuously perched, singly or in pairs, on the very top of some organ pipe cactus, sahuaro, or other tree where it can view the surrounding country; or occasionally it may be observed calmly sitting on the ground by the roadside, having then much

* The specific name, *pecten-aboriginum,* is a Latin-derived compound meaning liteally "comb of the aborigines," given in reference to the use the Indians made of the large burr-like fruit in combing the hair.

the appearance of a stuffed bird. In general it looks like a long-legged, long-necked, blackish-brown hawk with a white, black-crested head, white neck and breast, and white tail with terminal band of black. In flight the legs are held parallel and conspicuously downward, and there is white showing on the wing tips. Caracaras sometimes nest in the giant cacti but more often in ironwood trees. They are bold, adept hunters, both in flight and on the ground, when foraging for lizards and snakes. The name caracara, by the way, is of South American Indian origin and was first given because of the strange hoarse cry of its Brazilian representatives. Caracaras seem to be excessively parasitized by both biting and sucking lice.

Incidentally, it is well to point out the important role that the turkey buzzard, euphoniously called *zapolote* by the Mexicans, plays in the economy of the desert, especially in relation to populated areas of desert Mexico. These birds comprise what we may call Nature's "sanitary squad," for the disposal of carrion. When traveling one can often detect the probable site of a Mexican village long before one approaches it by the sight of great numbers of these picturesque black birds wheeling in the sky ahead. As you approach the village, the birds are frequently seen not only in the air, but along the roadsides, where they feed on dead animals which the people have thrown out to them. The Mexican people rightly regard buzzards with great respect, and few persons would think of molesting or killing them; this has made the birds relatively tame. Fortunate is the traveler who happens to see, as sometimes he may, a thousand or more of these somber grotesque scavengers, appearing like creatures carried over from a past age, as they rest during the heat of the day on the sandy banks or bathe in the waters of some slow-moving, meandering desert stream, such as the Río Mayo.

South of Guaymas one meets in numbers the somewhat smaller black vulture (*Coragyps atratus*), with body of almost uniform dull black; even the bill and bare skin of the head and neck are black. A conspicuous pale whitish or silver-gray area shows up on the underside of the wings as the bird is seen in flight. The tail is broad and very short. This bird, like its red-headed cousin, the turkey buzzard, is seen in great numbers about lowland villages and roadsides, where it feeds on carrion.

Among the strange animals of this section of the Sonoran Desert is the Sonoran collared peccary or javelina (*Pecari angulatus sonoriensis*). It is a relatively small salt-and-pepper–colored animal, unmistakably pig-like with long blunt snout, short pointed ears, and abortive tail. There is an erect mane of long black-tipped bristles, on neck, shoulders, and back, and a "collar" consisting of a narrow shoulder-band of buffy white. Peccaries hide during the middle of the day among cacti and other shrubbery, but at morning or evening small droves of them may be seen feeding in the open on "anything and everything"—from prickly pear fruits to roots, insect larvae, and snakes. They are swift, nimbly-running little beasts leaving behind them a strong musty odor.

The large desert mule or burro deer (*Odocoileus hemionus eremicus*), the "cuervo" of the Mexicans, inhabits the desert uplands on both sides of the International Boundary, even sometimes ranging a short distance up along the Colorado River, where it finds cover in the tree thickets. Like most mammals of the desert it is remarkable because of the extreme pallor of its coloration. It is called burro deer because of its large ears.

Other representative mammals of the area are the hooded skunk (*Mephitis macroura,* 132), badger (*Taxidea taxus,* 129), kit fox (*Vulpes macrotis*), Mearns' coyote (*Canis latrans mearnsi*), Arizona cottontail (*Sylvilagus audubonii arizonae*), the white-tailed, long-legged antelope jack rabbit (*Lepus alleni*), white-throated wood rat (*Neotoma albigula*), and cactus mouse (*Peromyscus eremicus*).

The rattlesnakes are certainly well represented in Sonora, nine species or subspecies being known. Most of them are kinds found also in the deserts of the United States. Most of the other Sonoran Desert snakes are similarly represented in arid portions of Arizona and in the Colorado Desert of California.

Many of the lizards found in southern Arizona and the Colorado Desert of California are common inhabitants of similar environments in Sonora. The Sonoran banded gecko (*Coleonyx variegatus sonoriensis*), characterized by its wide transverse body bands and mid-dorsal light line, can be seen in active state only at night; by day it hides under rocks. Among the Mexicans, who often think that this delicate, gentle, and harmless "squeaking lizard" is a young Gila monster, there is widespread fear of it. "Why," they say, "the salamanquesas are so poisonous that if one goes across a piece of firewood and a person picks up the wood, his hand may rot away!"

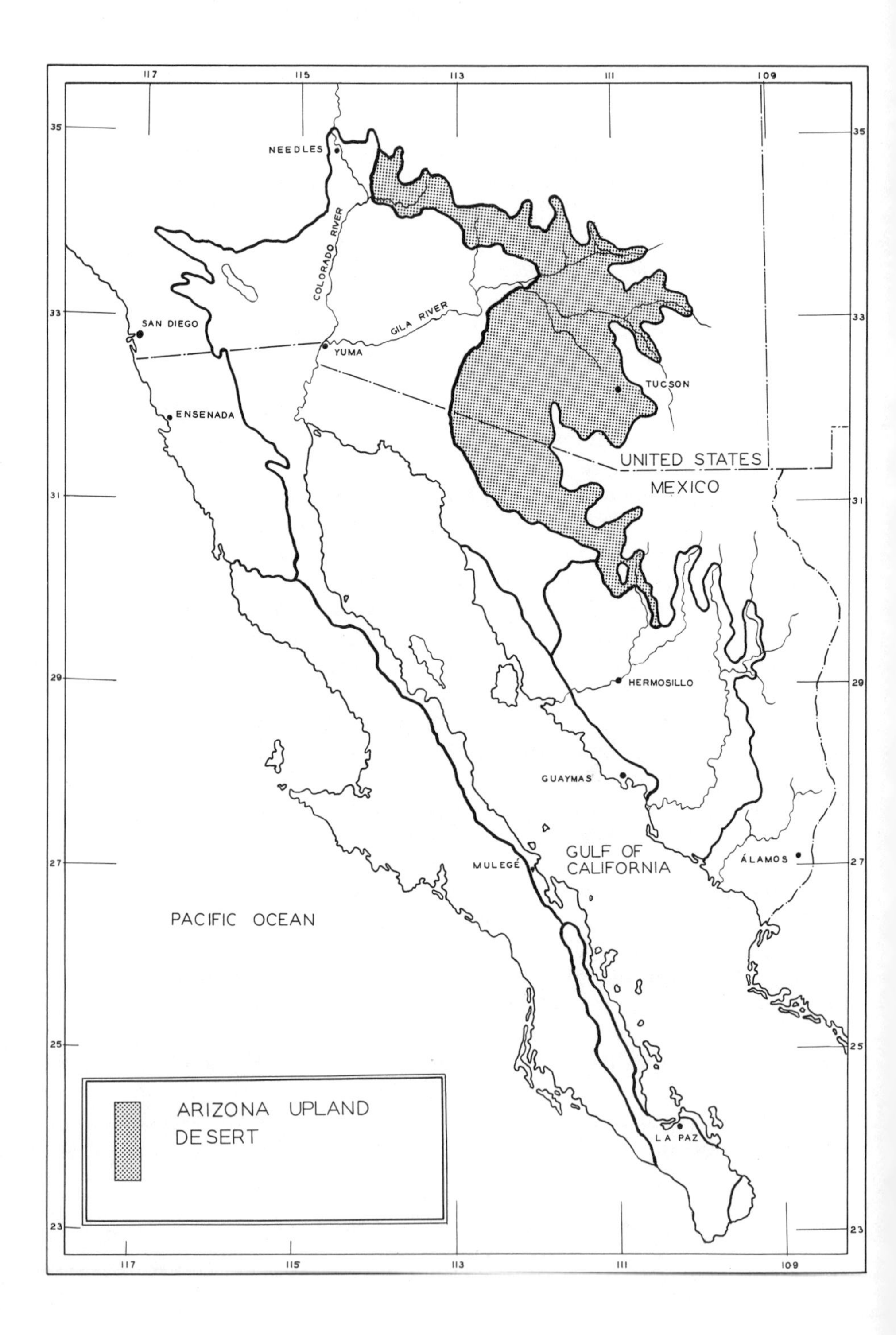
NEEDLES
COLORADO RIVER
GILA RIVER
SAN DIEGO
YUMA
TUCSON
ENSENADA
UNITED STATES
MEXICO
HERMOSILLO
GUAYMAS
GULF OF CALIFORNIA
MULEGÉ
ÁLAMOS
PACIFIC OCEAN
LA PAZ
ARIZONA UPLAND DESERT
117
115
113
111
109
35
33
31
29
27
25
23

6

THE *Arizona Upland* OR *Sahuaro* DESERT

The northeastern portion of the Sonoran Desert, as it occurs in Arizona, has no accepted name except the ARIZONA UPLAND DESERT, but since it is the heart of the domain of the Sahuaro cactus, it may well be designated as the SAHUARO DESERT. Included in it are the high deserts around Ajo, Tucson, Globe, and Wickenberg, and the rough, rocky, hill-studded region of north central Sonora around the Mexican towns of Altar, Magdalena, and Sonoyta. It is largely drained by the Bill Williams, the Gila, the Santa Cruz, the San Pedro, and the Sonoyta rivers. A large part of it is covered with small granitic hills or red and rugged volcanic mountains. There are no large stretches of sand, but in its western section are extensive flat arid valleys of gigantic measure. There are two rainy seasons, one in winter and one in summer. Because of the dual rainy periods, there are also two sets of ephemeral or short-lived plants. The late winter wild flowers are quite different from those which crowd the ground after midsummer showers.

"Never by any chance," wrote Dr. Forrest Shreve, "does a summer species appear in the winter or vice versa. In the warm moist soil of summer lie the seeds of the winter ephemerals as dormant as if they were in a dry refrigerator. This behavior is due solely to a difference in the temperature required for germination of the summer and winter plants. It is an easy matter to secure seedlings of winter ephemerals in summer by appropriate cooling of the soil."

The little-leaf or hill palo verde (*Cercidium microphyllum,* 223) is the most characteristic and often the most abundant of the larger plants of the rocky hillsides, and when it breaks into flower the whole land is brightened by its brilliant yellow blooms. The deep-rooted honey-bean mesquite (*Prosopis juliflora,* 218) is widespread over this desert and sometimes occurs in dense orchard-like thickets or bosques, especially where the water table is shallow; in less favorable situations it is a mere shrub. Other dominant plants are the creosote bush (*Larrea divaricata*, 238), the giant cactus or sahuaro (*Cereus giganteus*, 159), the biznaga or barrel cactus (*Echinocactus wislizeni*), the desert ironwood (*Olneya tesota*), the "tisito" (*Acacia constricta*, 226), bearing globular flower clusters of remarkable fragrance, the brittle bush or incense bush (*Encelia farinosa*, 267), the ocotillo (*Fouquieria splendens*), the goat-nut (*Simmondsia chinensis*, 257), the crucifixion thorn (*Holacanthus emoryi*, 336), and the burro bush (*Franseria dumosa*, 251). Perennial grasses are everywhere in evidence. The lechuguilla (*Agave lechuguilla*), so typical of parts of New Mexico and Mexico in the near-by Chihuahuan Desert, is strangely absent in Arizona, its place being taken by the valuable fiber-bearing bear grass (*Nolina*) and other kinds of yuccas and agaves, the latter known as century plants.

Among the handsomest and most widespread of the yuccas growing here is the palmilla (*Yucca elata*, 161). In some of the mesa lands and broad grass-filled valleys it is sometimes so plentiful that it forms thickets almost approaching the density of a forest. The slender trunk, simple, or as often branched, is sometimes 15 feet high. The numerous long, very slender leaves are white margined and have fine thread-like fibers along the edge. This is the common "soapweed" of southeastern Arizona and southern New Mexico.

In many portions of the Sahuaro Desert, especially on the mesas and well-drained bajadas, many kinds of brilliant-flowered cacti are unusually common, often in stands so nearly pure that they form veritable elfin forests. Some of the larger kinds, because of their peculiar appearance, are rather easy to identify. Some of these readily recognizable species are the upright-stemmed Teddy bear cholla (*Opuntia bigelovii*, 199), with its gleaming thick-set armor of golden yellow or greenish white needles that turn brown-black with age; the chain-fruited cholla (*Opuntia fulgida*, 300), which has a candelabra-like form and often long, curiously joined, and drooping chains of fruit; the buckhorn cactus (*Opuntia acanthocarpa*) with branching much like that of a deer's antlers; and several kinds of darning-needle or pencil cacti (*Opuntia*) with finger-sized stem-joints and very long, rigid gleaming spines. Then, too, there are numerous low-growing nipple cacti (*Mammillaria*, 351) and many small flat-jointed prickly pear cacti

(*Opuntia*). Most of the lesser-known kinds can be identified with certainty only with the aid of such books as Lyman Benson's *Cacti of Arizona* or W. Taylor Marshall's *Arizona's Cactuses.* The famed night-blooming cereus (*Cereus greggii*), with small, conspicuously ribbed stems that often resemble dried sticks, occurs chiefly, in the wild state, under trees and among shrubs. Its large white, handsome, and sweet-odored flowers open just after darkness and close up in the early morning. The stems arise from an often enormous, turnip-shaped root, which may weigh up to 60 or even 125 pounds!

Of cacti, the most remarkable one of this and the immediately adjacent desert region to the west is the giant cactus (*Cereus giganteus,* 159) or sahuaro—sometimes spelled saguaro. Speaking of its stem's construction, the desert botanist Volney M. Spaulding wrote: "The giant cactus, rising as it often does to a height of 50 feet in the form of a gigantic fluted column which may be simple or branched, is, mechanically speaking, a huge reservoir of water, subjected to the stress of high winds, and so constructed that for a long period of years it not only maintains securely its erect position and steadily continues its growth but also promptly expands whenever the soil is wet, even for a short time. The construction of such a tank represents an engineering feat which probably has no parallel in any artificial structure in existence, and the case appears still more remarkable when it is considered that the whole system of storage in an adjustable tank is dependent for its highest efficiency on the peculiarities of its root system." This root system consists of a strong anchoring taproot and thick secondary laterals in addition to the numerous, very long superficial rootlets, serving mostly as organs of absorption.

The fruits of many of the cacti, but especially those of the sahuaro, are harvested each season. Long hooked sticks are employed in jarring loose the sahuaro "apples" from the tall branches. The fruit pulp is eaten raw or stewed, and sometimes is permitted to ferment to form an intoxicating drink. Garcés, at the time of his travels in 1775, found the Indians much addicted to the drink. The seeds they crushed and used as flour.

The best showing of wild flowers of this part of the Sonoran Desert comes during the period spanned by the days of early March to mid-April, but, when they occur, the summer displays are also plentiful and vie with the spring flowers for beauty. The predominant spring blossom colors are yellow and white. Peculiarly, a large number of the annual plants, which spring up as if by magic after the winter rains, are identical with those found as spring flowers on the Mohave Desert and in the desert area of west Texas.

Grasses, perplexingly numerous in species and of great forage value and beauty, often are in bloom over great areas after the summer rains. Among the important kinds are several species of grama grass (*Bouteloua*), several species of tobosa (*Hilaria*), and sacaton (*Sporobolus*).

The beautiful, ubiquitous, drought-resistant creosote bush (*Larrea,* 230) often forms almost pure stands on the gently sloping alluvial fans, coloring these and the rocky slopes and hillsides yellow with its multitudes of small windmill-like flowers.

The white-throated wood rat (*Neotoma albigula*), rock pocket-mouse (*Perognathus intermedius*), cactus mouse (*Peromyscus eremicus*), ringtail cat (*Bassariscus astutus,* 124), kit fox (*Vulpes macrotis,* 125), and Arizona cottontail (*Sylvilagus audubonii,* 115) are characteristic mammals of this southern Arizona Upland Desert. Among the numerous birds one sees are the Arizona crested flycatcher (*Myiarchus tyrannulus,* 62); Gila woodpecker (*Melanerpes uropygialis,* 76), which drills nesting holes in the sahuaro trunks; gilded flicker (*Colaptes chrysoides mearnsi,* 77), also closely associated with the sahuaro; Palmer thrasher (*Toxostoma curvirostre palmeri*); cactus wren (*Heleodytes brunneicapillus couesi,* 99); western red-tailed hawk (*Buteo borealis,* 68); Gambel or desert quail (*Lophortyx gambeli,* 73); western turkey vulture or buzzard (*Cathartes aura teter,* 69); and elf owl (*Micropallas whitneyi*), which is the smallest of all owls, being near the size of a sparrow. The elf owl is closely associated with the giant cactus. Its series of rapid high-pitched notes is heard mostly at night. A bird certain to attract attention because of its peculiar and somewhat globular stick-nests is the small, peppery tempered verdin (*Auriparus flaviceps,* 101).

The Gila River, called the "Río del Nombre de Jesús" by Juan de Oñate in 1604 and "Río Grande de Gila" by Father Kino in 1694, is one of the major streams of this southwest desert area. Rising in the mountains of western New Mexico, it flows 500 miles across all of Arizona, emptying into the Colorado River just to the north of Yuma. It was until the time of the Gadsden Purchase (1853) a part of the International Border between the United States and Mexico. The desert area south of the Gila has long been occupied by the sedentary Pima, Papago, and Yuma Indians. The heroic German Jesuit, Eusebio Francisco Kino, spent almost twenty years in constant hazardous travels, often alone, among these and other Indians, going as far west as the Colorado River. In 1700 Kino descended the Gila River to its junction with the Colorado, and in the following year crossed the Colorado on a raft. It was the observations made on these exploratory journeys that convinced him that California was not an island. Kino found the Indians almost uniformly friendly and helpful.

There are several accessible and especially beautiful scenic areas in this portion of the Sonoran Desert. Very typical stands of the sahuaro, in a setting unspoiled by the press of civilization, may be seen in Sahuaro National Monument, fifteen miles east of Tucson, Arizona. This is an area of nearly 54,000 acres which was set aside in 1933 for the preservation of the best "forest" of this massive and noble tree, the state flower of Arizona. At the Monument headquarters is a well-laid-out and labeled cactus garden and an outdoor museum, where most of the desert plants common to the area can be readily identified. A circling drive leads the visitor leisurely through the heart of the Sahuaro forest, with new and often surprisingly beautiful views around every bend.

Another informing and delightful trip into this Arizona Upland portion of the Sonoran Desert, particularly if taken in March or early April, is the one over the paved highway from Tucson, by way of Sells, to and through Organ Pipe Cactus National Monument. (If you are traveling east from California you can reach the Monument by going south from Gila Bend to Ajo.) The country traversed in getting to this unique national monument from the east has long been occupied by the peace-loving Papago Indians.* Today the Papago, or "Desert People" as they call themselves, are a semiagricultural tribe, engaged in farming along with cattle raising and other related work. Many of them also work in the mines at Ajo from time to time. They now depend upon

* "It is of interest to note that the prehistoric Papago was a farmer and derives his designation from this fact. The characteristic crop plant was the native bean called pah', or, in the plural, pahpah', and the same term was applied to the tribe by neighboring peoples. The Spaniards slightly corrupted the appellation, pronouncing it Pahpaho' (the final vowel feeble and obscure), and spelling it, with some emphasis of the aspirate, Papago; the Americans retained this orthography, but pretty effectually concealed the original form of the tribal name by adopting the pronunciation indicated by their own orthoëpy. The tribesmen long ago accepted the name by which they were known among other tribes, adding the descriptive term a'atam, literally Beansmen, i.e., the Bean People." (W J McGee, ethnologist, quoted in *Desert Botanical Laboratory of the Carnegie Institution* [Washington, 1903], p. 15.)

wild fruits and seeds only as a supplement to their diet. In times past they successfully coped with severest desert hardships, and had learned to use many of the wild plants as sources of materials and of food and medicine. They made a tasty and nutritious porridge from the crushed seeds of the white-thorn acacia (*Acacia constricta*, 226), the honey-bean mesquite (*Prosopis juliflora glandulosa*, 218), and the palo verde (*Cercidium*, 223).

The land of the Papago is called Papaguería. It is a region of strange sounding place-names. Particularly are they strange when pronounced by the gutturally voiced natives:

Kom vo .... hackberry charco (Sp. *charco*, pond)
Gu achi .... big ridge
Pitoikam .... sycamore place
Topawa .... it is a bean
Schuchk .... black things
Stoa pitk .... white clay
Wahak hotrontk .... road goes down

The above are names found on little wooden signs along the roadsides, pointing to some of the 168 small ranches or villages scattered throughout the 3,000,000 acres of the Papago reservation.

Close eastern neighbors of the Papago are the Pima Indians, who once constructed substantial adobe structures but now live in dome-shaped shelters made of pliable poles covered with thatch and mud. Like the Papago they are of Shoshonean stock. The Pima are more numerous in Mexico, where they dwell along the Yaquí River in Sonora and to the eastward in Chihuahua. These Mexican Pima are now known as the Pima Baja or Nevome. Pima signifies "no" in the dialect of these Indians.

Some of the finest displays of wild flowers are seen where the road between Tucson and Organ Pipe Cactus National Monument passes in several places through rocky hills. Cacti of several kinds are there more abundant, as are also many flowering shrubs and perennial herbs and grasses of unusual beauty that cannot be found on the lower plains. In the Organ Pipe Cactus National Monument and surrounding Papago Indian Reservation are to be seen native specimens of the "pitahaya dulce" or organ pipe cactus (*Lemairocereus thurberi*, 149) and of the senita or old man cactus (*Lophocereus schottii*, 301), the only place they are found growing wild north of the Mexican boundary. Both are common farther south in Sonora and Baja California. South of the Monument headquarters are a short circular drive and a nature trail, with novel scenery and distinctive plants not to be observed readily elsewhere in the United States. The curious so-called Mexican jumping bean plant (*Sapium biloculare,* 143) can be examined growing here in its natural environment. It is the larva of a moth (*Carpocapsa saltitans*) inside the bean that causes it to jump and hop about. "This plant," said Paul C. Standley in his *Trees and Shrubs of Mexico,* "like *Sesbania pavoniana,* produces jumping beans. The juice is poisonous as in other species [of *Sapium*] and in Baja California the finely chopped branches are thrown in water to stupefy fish. Exposure to smoke from the burning wood or sleeping in the shade of the tree is said to cause sore eyes."

Going to Phoenix by way of Winkleman, Oracle, and Florence one passes through particularly fine, mixed stands of beautiful grasses, chain cacti, mesquite thickets, and sahuaros and, in the higher elevations, forests of high-desert shrubs and pines — all against backgrounds of Arizona's unbelievably blue skies and colorful volcanic mountains. Near Superior is the Southwestern Arboretum with its highly interesting collection of desert trees and shrubs. Near Mesa, the Desert Botanical Garden has trailside exhibits of desert flowers, trees, and shrubs.

West of Tucson in the low Tucson Mountains is a remarkable 30,000-acre sanctuary known as the Arizona-Sonora Desert Museum. It was opened in 1952. Here is one of the largest, most beautiful, and healthiest collections of sahuaro in the United States. In their midst are the Museum buildings, situated in a position commanding a magnificent view. Near by is a ten-acre "nature plot" where most of the native desert plants, including many desert trees and shrubs and other flowering plants of both Arizona and Sonora, are in view. Running through the garden are nature trails where labeled plants and various unique interpretive devices aid the visitor to gain knowledge of many desert plants. In addition there are large earth-colored pits containing the larger desert animals such as the peccary, the bobcat, and other wilderness inhabitants. Notable too is the display of living desert insects and spiders and other arthropods. A geology room includes a series of dioramas which graphically tell the "story of water." "The objective," said William H. Carr, the Museum's first director, "is to concentrate living wild life where visitors may examine them carefully at length and in a few hours obtain greater knowledge and understanding of desert fauna and flora than in many years of journeying." A considerable part of that remarkable and fascinating film, *The Living Desert,* was photographed in the Museum and its near-by surroundings.

A trip rich in scenic splendor is one in southern Arizona to the high and steep Baboquivari Mountains, which lie southeast of Tucson and near the Mexican boundary. The highest elevation of the bold range is Baboquivari Peak, 7,740 feet. The Papago name for the culminating thumb-shaped pinnacle means "with bill or beak in the air," given in reference to the "eagle-headed peak." Here there are occasionally to be observed a number of rare birds that have wandered up from Mexico and are found nowhere else in the United States, such as the thick-billed parrot (*Rhynchopsitta pachyrhyncha*).

7

THE *Sonoran* DESERT: WESTERN SUBDIVISIONS

The largest and most arid part of the Sonoran Desert lies along the lower Colorado River and upper part of the Gulf of California. Here the rainfall is scanty, and the days of bright sunshine, dessicating wind, and high temperatures are indeed great in number. The surface features of this region include almost every kind of terrain from sandy or gravelly plains to alkali flats and picturesque low mountains of volcanic and granitic origin.

This expansive area includes (*a*) almost all of low-lying southwestern Arizona and the great stretch of sand desert in northwestern Sonora that lies between the foothills and the Gulf of California; (*b*) the Salton Sink, together with the region immediately to the east of it along the Colorado River in California and the narrow strip of Baja California coast along the upper western side of the Gulf of California as far south as Bahía de Los Angeles; and (*c*) the central portions of the peninsula of Baja California comprising the vast Vizcaíno-Magdalena Desert with its unusually strange flora. Runoff waters from all these areas eventually drain by way of intermittent streams, washes, or the few permanent rivers (such as the Gila and the Colorado) into the Gulf of California, the Salton Sea, the below-sea-level basin occupied by the Laguna Salada in Mexico, or the Pacific Ocean.

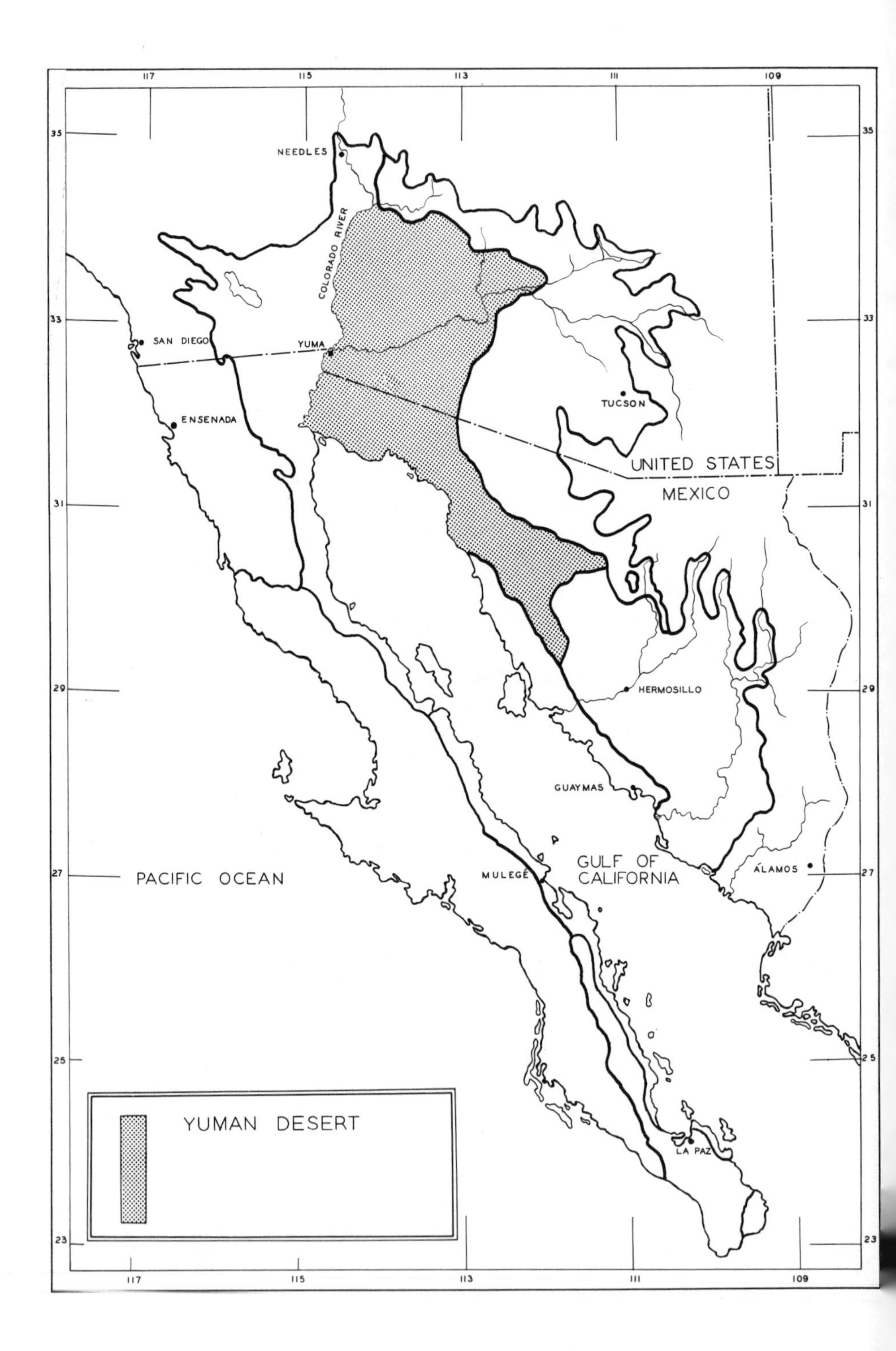

NEEDLES
COLORADO RIVER
SAN DIEGO
YUMA
ENSENADA
TUCSON
UNITED STATES
MEXICO
HERMOSILLO
GUAYMAS
MULEGÉ
GULF OF CALIFORNIA
ÁLAMOS
PACIFIC OCEAN
LA PAZ
YUMAN DESERT
117
115
113
111
109
35
33
31
29
27
25
23

8

THE *Yuman* DESERT

West of the southern part of the Arizona Upland Desert and extending almost to the Colorado River is a portion of the low-lying Sonoran Desert known as the YUMAN DESERT. Southward it takes in also the sandy western plains of northwestern Sonora between the foothills and the Gulf; northwestward a narrow strip extends near the Colorado River to as far as the Bill William River and to the north of Needles, California. Here is an arid land indeed, with an average rainfall of only about 4 inches a year. For the most part it consists of vast low sandy plains, extensive dunes, scattered hills of highly eroded volcanic rock, and low barren boulder-strewn mountain ranges of granite such as the Cabeza Prieta, the Mohawk, and the Gila mountains, which run parallel to one another with a general north-south trend. Much of the Yuman Desert lies within the tract originally acquired from Mexico in the Gadsden Purchase of 1854.

Among the most common shrubs are the ever-present creosote bush (*Larrea*, 238), two types of burro bush (*Franseria dumosa*, 251, and *F. deltoidea*, 249), and the very widespread brittle bush (*Encelia farinosa*, 267). Graceful desert "willows" (*Chilopsis linearis*, 297), palo verdes (*Cercidium floridum*, 227, and *C. microphyllum*, 223), and ironwoods (*Olneya tesota*, 355) are plentiful along the margins of many of the dry watercourses; thorny smoke trees (*Dalea spinosa*, 348) are common in the sandy wash-bottoms. The distinctive sahuaro cactus (*Cereus giganteus*, 159) is found on the well-drained bajadas, as are also the ocotillo (*Fouquieria splendens*), gray thorn (*Condalia lycoides*, 144), and cat's-claw (*Acacia greggii*, 221). Oddly enough, except in a few places, yuccas of any kind are almost entirely absent from the country south of the lower Gila River. The small torote or elephant tree (*Bursera microphylla*, 293), with its corpulent reddish-brown and papery-barked trunk, its rubbery branches and tiny compound leaves, grows on rocky foothill slopes mostly from 100 to 2,000 feet in elevation; it is a tree of wide distribution in adjacent Sonora and Baja California.

The creosote bush (*Larrea divaricata,* 238) is, except in the Great Basin and Navahoan deserts, the most widespread and common of all our desert shrubs and is certainly typical of the entire Sonoran Desert. It is believed to have originated in the deserts of South America. Often it covers extensive areas, miles in extent, in pure stands or in company with the burro bush (*Franseria,* 251). It is remarkable for its ability to withstand protracted periods of extreme drought and for its subsequent rapid comeback after only slight penetration of the soil by rain. The deep green of its glutinous, strong-scented foliage gives a characteristic color to much of the desert landscape. When wet the leaves fill the air with a pleasant memory-provoking odor; no other scent is more reminiscent of the desert. The wood when burned in the campfire gives off a hot flame and an aroma that is always pleasing. From the leaves can be extracted an acid which is useful as an antioxidant and which prevents or at least retards fatty foods from turning rancid. Because of the oily appearance of the new leaves, this shrub is often erroneously locally called greasewood (see page 143). Flowering of the creosote bush is most profuse in spring but may take place at any time during the year if there is sufficient rain to induce new growth.

It is interesting to note how, on this desert particularly, the distribution of shrubs is favorably affected by the waters only intermittently flowing in the small, ordinarily dry watercourses. Here there is enough extra moisture, even though the actual amount is seldom great, to encourage the growth of hardy species of dwarf woody plants. If one climbs the mountains and looks down upon the broad plains and lower mountain slopes, he will see the intricate patterns of the numerous small drainways marked by the crooked lines of hardy gray-green shrubs and occasional trees which have grown up along them. Between these shrub-lined watercourses are broad barren areas which are covered with numerous flat stones, which have dropped lower and lower as the loose material has been blown or washed away from between or beneath them. They have finally become closely wedged together and in some cases partially cemented in place by minerals which rain has dissolved from the soil about them. The end result is what geologists term a "desert pavement," and in many places, over wide areas, it looks exactly as if man had cleared these spaces and deliberately created a marvelously flat and even mosaic of stones.

If, instead of stones, the surface of the ground is fine gravel or clay, then the sparsely shrubbed ground may have a binding cover or crust of peculiar black ground lichens (*Acarospora* and *Lecidia*), so that it resembles a vast carpet of blackish felt. Such lichen-covered soils are particularly noticeable on the near-level basal slopes surrounding the Gila and Mohawk mountains southeast of Yuma, Arizona.

In the low, bleak desert ranges just east of the Colorado River in Yuma and Mohave counties of Arizona grows a broad-leafed bear grass (*Nolina microcarpa*). It has a robust trunk sometimes five or six feet tall. In form it somewhat resembles a yucca, but has more pliant and grass-like leaves, much smaller and less showy flowers, and rounded rather than flat seeds. This bear grass often grows in groups or clumps which from a distance may resemble small palms.

There are four near-trees or large shrubs of this division of the Sonoran Desert which are outstanding because of their numerous, rigid thorny branches and because of an almost complete lack of functional leaves. These shrubs are the canotia (*Canotia holacantha*), crucifixion thorn (*Holacantha emoryi,* 336), junco (*Koeberlinia spinosa,* 147), and the smoke tree (*Dalea spinosa,* 348). In these plants the work ordinarily performed by the leaves is taken over by the yellowish or gray-green bark of the persistent stems and branches. The canotia and junco are found often in dense thickets on rocky slopes and hills. Crucifixion thorn generally grows on desert plains or along the edge of clay pans. The smoke tree, with higher water requirements, is wholly confined to the bottoms of sandy washes. Of these four shrubs only the smoke tree may be said to be a really handsome plant, even though it is only an intricate mass of sharp spines. However, these spines are comparatively small and of such a soft, beautiful, bluish gray-green color that the whole tree when seen from a distance looks quite like an ascending column of blue smoke. The season of flowering for the smoke tree is in April or May; then it becomes a mass of rich indigo blue flowers—as fine a sight as one could wish to behold.

Probably the hardiest of all desert trees is the desert ironwood (*Olneya tesota,* 355), which grows abundantly along dry stream beds in most parts of the Sonoran plains and on the low, hot desert land of southwestern Arizona and adjacent eastern California, and in Baja California to the tip of the peninsula. The brittle wood is so hard that it soon dulls the sharpest tools; for this reason the

Mexicans have called it palo de cierro ("tree of iron"). The wood has little commercial use except as fuel. During very dry periods the tree may lose practically all of its leaves, but ordinarily it is covered with gray-green foliage, offering welcome shade to the summer camper. In late spring the tree breaks into flower. The pea-like blossoms, borne in short racemes, are purple and white, and are followed by short several-seeded pods which hang on until near autumn. The hard dark-brown seeds form an important item of diet for many desert animals. The tree may reach a height of 30 feet, but the average is from 12 to 20 feet with a spread of 15 to 20 feet. Although it rejoices in heat, the desert ironwood is quite intolerant of frost.

In the ephemeral spring wild flower season, the broad spectacular fields of apricot-colored desert mallow (*Sphaeralcea,* 282) are certain to arrest the attention of lovers of great masses of brilliant color. Some of the largest of these mallow "gardens" can be seen in the sandy soils along the road between Gila Bend and Yuma and also between Gila Bend and Ajo, in Arizona. Other very abundant roadside spring flowers, mostly annuals, are the desert marigold (*Baileya pauciradiata pleniradiata,* 276), with large disk-like yellow flower-heads and silver-green hairy leaves; the bright yellow short-statured sunflower (*Helianthus canus,* 270); the showy white-flowered prickly poppy (*Argemone platyceras,* 170), so plentiful along Sonoran roadsides; and the purple-pink sand verbena (*Abronia villosa,* 287). The bluebonnet (*Lupinus sparsiflorus*) is particularly beautiful when occurring in pure stands. This is also true of the desert sunflower (*Geraea canescens,* 278) and the various evening primroses, especially the small-flowered varieties such as the yellow *Oenothera brevipes* (243) and the white *Oenothera scapoides.* The sweet-scented "dune primrose" (*Oenothera deltoides*), particularly plentiful in the open sandy areas, has big white papery flowers which unfold near sundown and remain open until late morning. When blooming, the blossom-bearing branches of this showy plant lie almost flat on the sand, radiating from a central root-crown. Later, as the "primrose" goes to seed, these woody stems curl upward over the central, "thorny" main stem and form what is called locally a desert birdcage or primrose basket. The white-flowered desert "carnation" (*Chaenactis fremontii,* 188) and purple-flowered "heliotrope" (*Phacelia crenulata,* 321) when occurring in masses add much beauty to the floral show occurring along desert highways.

Across the lonely and all but waterless plains of northwestern

Sonora and southwestern Arizona traveled the courageous Father Kino and his retinue of followers, late in the eighteenth century. They made a trail from Kino's headquarters at Sonoyta, in northern Sonora, to the Colorado River, along a route that was later to be aptly termed the Camino del Diablo (Sp., "Road of the Devil") because of the many travelers who perished along it from heat, hunger, and thirst. One of the Camino's few dependable "water holes" was at Tinajas Altas ("high tanks"), located about 30 miles south of Welton in Arizona, on the east face of the Gila Mountains, a picturesque narrow granitic range which parallels the Río Colorado. These "tanks," eight in number, are water- and gravel-eroded cavities worn out of the solid granite in a steep arroyo. They are from time to time filled with rain water, which often remains for long periods. Not all of these natural storage tanks are reliable sources of water, especially in midsummer, and those that are the most likely to contain water are located high up and are difficult of access. Passing travelers and the Mexican bighorn sheep (*Ovis cremnobates,* 134) still drink from these accumulations of soft water.

Some of the most beautiful stretches of Sonoran Desert are found in the vicinity of the Tinajas Altas and the Mohawk Mountains to the east, but very poor roads offer little inducement for the average tourist to see them. This particular region is the home of Mearns' coyote (*Canis mearnsi*), named in honor of Dr. Edgar A. Mearns, who traversed this area as a naturalist and physician to the U.S.-Mexican Boundary Survey party from 1892 to 1894. Here, too, occurs in large colonies the low-growing, yellow-flowered Kunze cholla (*Opuntia kunzei*), with sharp, two-edged, dagger-like spines set on the tubercles of short-club-shaped joints.

To the north of Tinajas Altas, along the highway between Yuma and Quartzite in Arizona, an almost equally attractive country of similar terrain may be seen without undue effort. To the east one can see the magnificent Castle Dome and Kofa mountains, both of volcanic origin and consisting of inspiring pinnacles, steep-faced crags, slot-like ravines of great magnitude, and plugs of rhyolite stripped bare by erosion. The name for the Kofa Mountains was obtained by combining the initials of the name of a large mine in the range, the "King of Arizona." Far back and high up in one of the secluded high-walled and narrow gorges of the Kofa Mountains is hidden the only native group of desert fan palms (*Washingtonia filifera*) to be found in Arizona.

Days and nights spent in camping and exploring in this land

of bewitching scenery can never be forgotten. It is Arizona's desert at its very best. At evening time the volcanic cliffs and spires are glorious with sunset hues; the early morning skies are clean and brilliant.

South by way of Ajo and Organ Pipe National Monument to the border town of Sonoyta, Mexico, and then over a good paved road to the small fishing village of Punta Penasco, some 65 miles below the boundary, there are many additional fine specimens of the organ pipe, sahuaro, and senita cacti, as well as splendid old ironwood trees growing in the washes that drain the low mountain ranges. In the distance to the northwest of Punta Penasco are the famous Pinacate Craters, more than a hundred in number; their broad lava beds are sometimes referred to as the "Sea of Broken Glass." Together they form an island of black and rich brown in the midst of a broad, near-white sandy plain bordering the Gulf of California. Pinacate was the first-described volcano in North America. In ancient times a great stream of lava pouring out from the Pinacate Craters diverted the Río Sonoyta from its former course so that it could no longer empty into the Gulf near the mouth of the Colorado River. Its waters are now absorbed into the desert sands near the eastern border of the lava. But side trips to this rugged Pinacate volcanic area are not to be recommended to the newcomer since the roads for the most part are unmarked and very rocky and sandy. Except occasionally in "tanks," there is no fresh water to be had. For that matter, going off the main road into any portion of this wild region of loose sand, volcanic ash, and tire-cutting lavas is somewhat risky even for the seasoned traveler who goes equipped for hardships.

Between the Gulf itself and the Pinacate area is a huge, largely unexplored area of sand called by Mexicans the "Gran Desierto." Forrest Shreve, the famous desert botanist, wrote of this wild and lonely territory: "The prevailing winds have piled great dunes against the west side of Pinacate and most of the smaller peaks. In many places sand has blown up on the western slopes of the hills for several hundred feet . . . and has completely buried the rocks. There is rarely enough rain to undo the work of the wind. Opposite every opening between the hills one or more long dunes stretch eastward, some of them extending 30 to 40 miles from the Gulf."

The small Mexican agricultural village of Sonoyta, just across the border, is as yet little spoiled and offers the visitor an opportunity to see firsthand the leisurely, peaceful life of a small bit of rural Mexico. Punta Penasco ("Rocky Point") is a mecca for deep-sea fishermen. It is well worth visiting if only to see the deep blue of the Gulf waters laving the ebony black cobblestone shore, or to see there the picturesque array of fishing boats of all sizes and designs crowded into the shallow bay. Further it is interesting to watch the unloading of the day's catch, whether it be the huge totuava, formerly caught just for their swim-bladders, which were dried and shipped to the Orient for soups, or the giant tortugas or sea turtles, which are harpooned. Close by is a beach called Bahía Cholla ("Cholla Bay").

For ages the swift-flowing Colorado River has been an effective western barrier for a number of Arizona's most interesting desert plants and animals. With the exception of a few stray specimens found just across the state line in California, from Parker Dam southward, the giant cactus or sahuaro stops abruptly at the river. This probably can be explained by the fact that the Colorado River pretty well marks the western limit of summer rains. The tiny elf owl (*Micropallas whitneyi*, 60) and the gilded flicker (*Colaptes chrysoides*, 77), which are almost wholly dependent upon the sahuaro for nesting sites, are plentiful in Arizona but only sparingly found on the California side of the river. The small Yuma antelope squirrel (*Citellus harrisi*) dwells in the rock desert almost to the river's edge in southern Arizona; in California, only a few hundred yards away, it is absent. On the other hand, the white-tailed antelope squirrel (*Ammospermophilus leucurus*, 118) is plentiful in California but not in southern Arizona. Two Arizona species of pocket mice (*Perognathus*) and one white-footed mouse (*Peromyscus*) are found in the Yuma district east of the Colorado River but do not occur farther west. The river also acts as a barrier to the westward migration of the javelina or peccary (*Pecari sonoriensis*, 130), of the only poisonous lizard in the United States, the Gila monster (*Heloderma suspectum*, 40), and of the very poisonous coral snake (*Micrurus euryxanthus*). All of these animals, with the exception of the peccary, are strictly associated with desert conditions and are not dependent upon streams or springs as sources of water; neither do they take to the water to swim. There are some rodents not so specialized, and some carnivores such as the bobcat, coyote, and skunk, which are identical on both sides of the Colòrado.

This Yuman district of the Sonoran Desert is the home of the gilded flicker (*Colaptes chrysoides*, 77) and diminutive elf owl (*Micropallas whitneyi*, 60), as well as the Gila woodpecker (*Mel-*

anerpes uropygialis) and the vigorous-voiced Arizona crested flycatcher (*Myiarchus tyrannulus,* 62), all associated with the sahuaro. The small gray-backed, white-breasted Lucy warbler (*Helminthophila luciae,* 59) is closely associated with the mesquite. The Nuttall poorwill (*Phalaenoptilus nuttalli,* 83) is heard throughout the night in spring, summer, and early autumn. Ravens (*Corvus corax*), Harris hawks (*Parabuteo unicinctus harrisi,* 71), and red-tailed hawks (*Buteo borealis,* 68) are to be seen in the sky of daylight hours.

Desert travelers will see along the road between Gila Bend and Yuma the little Harris antelope squirrel (*Citellus harrisi*), which except for its gray tail looks much like the white-tailed antelope squirrel of the adjacent Colorado and Mohave deserts in California. In sandy areas the small round-tailed ground squirrel (*Citellus tereticaudus,* 117), with its bleached gray body, big beady black eyes, exceedingly small ears, and pencil-sized tail, can be glimpsed darting across the highway or going about eating flowers and harvesting seeds. If you camp out in this scenic region, you may have the good fortune to have a little cat-size kit fox (*Vulpes macrotis,* 125), "all eyes and ears and tail," and always gentle and inquisitive, come right up to your camp. There it can be observed by the flickering light of the fire as it investigates your belongings or just sits quietly and stares at you in fascination.

Other animals sometimes seen by campfire light are the nervous Bailey pocket mouse (*Perognathus baileyi*), the very abundant desert pocket mouse (*Perognathus penicillatus*), the little spotted skunk (*Spilogale arizonae,* 133), and the big-eyed desert kangaroo rat (*Dipodomys deserti,* 120), distinguished by its large size and white-tipped tail. The white-throated pack rats (*Neotoma albigula,* 119) make large protective accumulations of miscellaneous available materials, including cholla cactus needle-clusters and joints, deep beneath which they dig an underground nesting site. From the base of "rubbish" radiate cleared passageways, permitting entrance and exit in all directions. Most of the nocturnal mammals just mentioned have paler-colored fur than similar species found in the less arid Tucson desert district farther east.

The trade or pack rat, so called because of its occasional habit of trading or packing off one object for another, may often be heard after dark as it forages about for food or busies itself with the cutting of fresh twigs to protect and adorn its nest, or with the gathering of debris to pile on top of its already bulky "house of rubbish." As a rule, only one pack rat lives in each nest pile, and ordinarily as soon as the young are mature they are driven from the immediate area and must seek new places to build their homes, or else take over an abandoned dwelling of some deceased pack rat. The same nest site may be occupied again and again by different pack rats over a period of many years, with each new occupant adding his layer to the stack of debris. In some cases, these nests may be as high as three feet, with a diameter of four or five feet! Like the Indians and early travelers, campers of today find these accumulations of dry sticks a ready source of fuel.

It may always be considered a moment of good fortune when, as so often happens, eleodes (the circus bug, *Eleodes* sp.) comes patiently crawling along into camp. This large black beetle stands on its head and absurdly kicks its feet in the air when poked with a stick. When in this angular position, or when crushed, some species of eleodes emit an offensive, musty-odored liquid and are therefore locally called stinkbugs. These strange insects are near-relatives of a group of dark-colored beetles found in semitropical deserts in many parts of the world. They feed mostly upon dry vegetation. Inoffensive and harmless, these ever-wandering insects always bring cheer to the solitary camper and are welcomed as familiar friends.

Doubtless the most grotesque animal of the Yuma district is the unique Gila monster (*Heloderma suspectum*, 40). This large lizard and its near relative, the Mexican beaded lizard, are the only saurians in the world that are definitely known to be poisonous to man. The generic name *Heloderma* means literally "nail-skin" or, more loosely, "nail-headed skin," in reference to the numerous big bead-like scales of black, gray, and salmon-pink which stud the body and tail. What its original describers thought of this reptile is revealed in the now obsolete Latin specific name "horridus." The clumsy body form, short chubby limbs, and usually obese blunt tail indicate the lizard's ordinarily lethargic habits, but on occasion in warm weather the Gila monster can move quite rapidly for short distances. Like many other desert lizards, it stores fat in its tail, and the condition of the lizard can be told at a glance by the amount of fat in its tail. The venom, almost as lethal as that of the rattlesnake, is secreted by a pair of large modified salivary glands under the lower jaw. The teeth or "fangs" are grooved but are not hollow like those of the rattlesnake. In order to effectively poison its prey, the Gila monster seizes it with bulldog grip and with persistence "chews" the poison into the flesh, sometimes turning on its back so that the venom can flow down into the wound more quickly. Folklore has it that this bizarre creature has no anus, and that it is poisonous because of accumulated body wastes, but this is sheer fantasy. The Gila monster is usually so sluggish and docile that it can be tamed and may be handled without fear; the chances of being bitten are exceedingly rare. Oddly enough, in captivity the Gila monster loves water and soaks in it long hours!

Sometimes one sees what appears to be a bit of bright red cotton moving across the surface of the ground. Closer inspection reveals the fact that it is an insect, looking much like a large ant nearly half an inch or more in length. It progresses very rapidly, pausing here and there, and then hurrying on, as if in search of something. This is the velvet "ant" (*Mutilla sp.*, 24), which is not an ant at all but a wingless wasp. The much smaller male mutillid has wings, but for her entire lifetime the female is wingless. She spends most of her time searching out subterranean bees' nests in which she may lay her eggs. Not all velvet ants are red in color; other species are silvery white or gray, and some are brilliant orange or yellow. When disturbed they emit a clearly audible squeaking noise, and have long stings which can produce a burning pain.

At other times, especially after a heavy rain and in dune country, the bit of red that can be seen slowly progressing over the sand may prove to be the dainty angellito, or velvet mite (*Trombidium magnificum*). The velvety appearance of the back of this jewel-like mite is due to a dense covering of short hair. During a portion of its life history the angellito lives in the sand quite out of view.

In the filled soils along the east side of the Colorado River above Laguna Dam, fossil wood may be found. The mineralized logs and tree fragments are not large or colorful but show grain and ring structure very clearly. The trees that were petrified are said to be closely related to hardwood species found growing on the desert at the present time.

The cottonwood-covered Colorado River bottoms about the mouth of the Gila River were long occupied by the Yuma Indians. They also lived on the west side of the Colorado River. To the south of them, nearer the Gulf of California, dwelled the Cocopa. In many ways the Yuma were much like the better-known Mohave who lived along the Colorado River to the north of them. The two tribes were virtually identical in their methods of agriculture, house-building, clothing, close-combat methods of warfare, and desire for renown in battle. The Yuma feuded much with the Maricopa people who lived farther east on the desert along the Gila River, and finally became embroiled with the peace-loving Pima, against whom they repeatedly fought. Their last great war against these sturdy farmers was in 1858; it ended disastrously for the Yuma when the Pima almost annihilated them. Fort Yuma was built in 1851 at about the center of Yuma territory to protect overland travelers from Indian menace while passing through on their way to the settlements and gold mines of California.

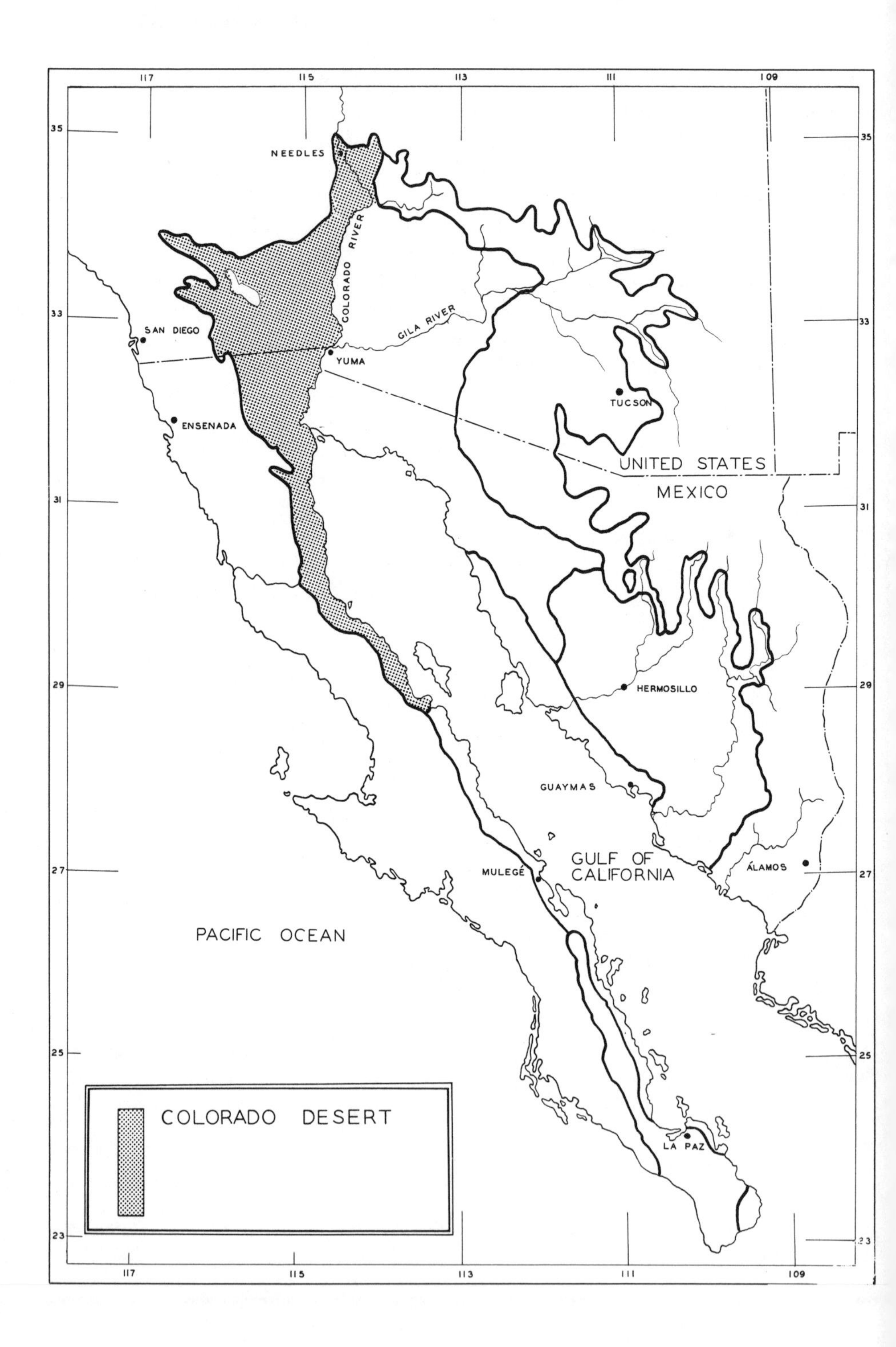
117
115
113
111
109
35
33
31
29
27
25
23
NEEDLES
COLORADO RIVER
GILA RIVER
SAN DIEGO
YUMA
ENSENADA
TUCSON
UNITED STATES
MEXICO
HERMOSILLO
GUAYMAS
GULF OF CALIFORNIA
MULEGÉ
ÁLAMOS
PACIFIC OCEAN
LA PAZ
COLORADO DESERT

9

THE *Colorado* DESERT

To the west of the Colorado River, mainly in southeastern California and northern Baja California, lies the COLORADO DESERT. Included in it are all of those areas which drain directly into the Colorado or into the Salton Sea, which from time to time in the past has had a direct connection to the river. Much of this desert lies below sea level, or just a little above it. The southern portion, south of the Salton Sea, in California, is called the Imperial Valley; like the Gulf's delta area south of the International Border, it has very deep and exceedingly rich alluvial soils which in recent years have been utilized more and more as agricultural lands, mainly for cotton and alfalfa growing and for truck farming. The portion of the Colorado Desert to the north of the Salton Sea, called the Coachella Valley, has equally rich soil, and today is the home of the California date industry. Most of the remainder of this region is sand and rock desert where the dominant plant cover consists of creosote bush (*Larrea divaricata,* 238), burro bush (*Franseria dumosa,* 251), and brittle bush (*Encelia farinosa*). Compared with the best of the arboreal or tree deserts of Sonora and southern Arizona, the Colorado Desert seems barren indeed.

Since to many the name "Colorado Desert" seems to be a poorly chosen one, it is well to point out that it was given because most of the area lies along the Colorado River and not because of any connection it has with the state of Colorado. This basin-like arid desert, euphoniously called "La Palma de la Mano de Díos" ("the hollow of God's hand") by the Mexicans, was named the Colorado Desert by William P. Blake in 1853, eight years before the state of Colorado was named.

In the broad sandy washes cutting the detrital slopes which border the steeply rising mountains, smoke trees (*Dalea spinosa,* 348), palo verdes (*Cercidium floridum,* 227), and desert willows (*Chilopsis linearis,* 297), so prevalent to the eastward, are still found, but the widely traveled desert student certainly notices the absence of the large cacti and many of the shrubs which are so common in southern Arizona. The desert ironwood (*Olneya tesota*, 355) is here, too; however, it is largely confined to the broad washes and gullies of the small, somewhat mountainous area between the Salton Sea and the Colorado River and southward along the Gulf of California.

So intimately is the history of the Colorado Desert linked with that of the lower Colorado River and the story of the Salton Sea, that a few statements concerning the geologic and recent history of the river and the sea are essential to a proper understanding of the region. Here there is indeed a close correlation between geology and biology.

The Salton Sink (also called the Salton Basin), wherein now lies the Salton Sea, is a depression some 273 feet below sea level at its lowest point and without a drainage outlet to the Gulf of California. It was formed by the gradual sinking of a 200-mile-long block of the earth's crust at a time when surrounding mountain ranges were slowly being elevated to the east and west. This "graben," as geologists term such a fault-depressed region, would now be nearly filled with water from the sea, at least as far north as Indio, if it were not for the natural fan-like "dam" or delta which the Colorado River built across its lower end where for long periods it discharged its tremendous load of silt into the upper end of the Gulf. Much of this material was derived from the Colorado Plateau far to the northeast, while the river was sculpturing the Grand Canyon. During heavy floods the river from time to time radically shifted its course across its low-lying delta and overflowed so that it sometimes emptied south into the Gulf and at other times northward into the below-sea-level Salton Sink or into another sink wherein now lies Laguna Salada. It created ephemeral shallow lakes in the sinks; these later evaporated, leaving the floor a dry flat until the next flooding.

Much more recently, but still many, many hundreds of years ago, the Colorado discharged into the Sink for a prolonged period, there to form a very large inland fresh-water sea with a depth of more than 300 feet. This sea was over 100 miles long and 35 miles across at its widest point. It extended from above the present town of Indio, California, to 17 miles south of the present U.S.–Mexican boundary. To this ancient body of water the geologists have given the name Lake LeConte (commemorating the able geologist J. N. LeConte, who early made studies of this sink); it has also been called Lake Cahuilla, a name given by William

L. Burr Belden

Abundant in many of the sandy washes of the low, hot Colorado Desert is the smoke tree (*Dalea spinosa*). In the sand are the holes of the large desert kangaroo rat.

Edwin Avery Field

Scene on the Yuman Desert, far western Arizona near the Colorado River. The large cactus is *Cereus giganteus*. The slender-stemmed shrub is the ocotillo (*Fouquieria splendens*). Deerhorn cactus (*Opuntia echinocarpa*) and burro bush in foreground.

Typical Colorado Desert scene with much-dissected mountains of igneous rocks, cactus-covered (*Opuntia bigelovii*) areas below. The rounded shrubs are burro b (*Franseria dumosa*), brittle bush (*Encelia farinosa*), and cat's-claw (*Acacia greggii*

Edward A. Hamilton

Kenneth Middleham

Eleodes, the tumblebug, is one of the most conspicuous large insects of the Sonoran Desert. It is a frequent camp visitor.

Typical of the Colorado Desert are its broad plains dotted with low shrubs and fine symmetrical specimens of the ocotillo (*Fouquieria splendens*), one of which is seen to the right of the road leading down a sandy wash. A dry lake, its flats encrusted with alkali lies at the base of the barren mountains.

P. Blake, who was among the first to do extensive research on the origin and nature of the entire Salton Sea area. The name now most preferred is Lake LeConte. The sharply defined prehistoric beach lines and wave-cut terraces of this old fresh-water lake are still very evident, especially along the base of the spurs of the Santa Rosa Mountains at the northwest end of the present Salton Sea. Fine views of them may be had to the west as you drive from Indio toward El Centro.

A stop at Travertine Point, which was once an offshore island in Lake LeConte, is well worth while, for here can be seen at close range the old beach line and the strange, porous "travertine" (really a calcareous tufa formed by minute blue-green algae which encrusted the rocks that were under water at that remote time). In some places the coral-like "travertine" has been incised by some unknown prehistoric Indians as they made curious petroglyph designs. Many of these markings appear to have been partially obscured by later deposits of tufa, perhaps indicating that the people encamped along this shore as the encrustations were being precipitated. Near by, along other old beach lines, are some odd circular walls of tufa-covered rocks which have been called "fish traps." Today there is no agreement among anthropologists as to whether these crude walls were used as fish traps or as foundations for brush houses, as hunting blinds or as ceremonial sites.

The first of the two ancient Lake LeContes was probably an Inter-Glacial or Post-Glacial fresh-water body, but it is obviously younger than the Pluvial lake chains of the Great Basin and Mohave regions (see pp. 125 and 146). Evidently it was in existence for a long time, perhaps several thousand years. This lake finally dried up after a southward shifting of the Colorado River's course; then followed a dry period of long duration. There is strong evidence of a second complete or high-level filling of Lake LeConte in recent times. Based on archaelogical findings, this lake had a duration of about 450 years—from about A.D. 1000 to 1450 or 1500. That one or more high stages of the lake were accompanied by an abundant aquatic life, especially small fresh-water mollusks, is attested by the numerous fossil shells which remain to this day. Sometimes they occur in such numbers that in places the wind-blown sand surfaces are white with them. In allusion to these small shells, the valley northwest of the Salton Sea was named Coachella (a misspelling of the Spanish "conchilla," little shell).

Fossil evidence in the form of marine clams and snails seems to indicate that the area at present occupied by the Salton Sink was at least twice covered by the sea. The first sea invasion was before the Salton Basin existed. Recent studies seem to show that the last covering by the sea may have taken place between the two fresh-water stages of Lake LeConte just described. It is thought that the sea waters may have come in when very high tides from the Gulf of California entered the Sink through the low trough found at the western edge of the Colorado River delta. This invasion of tidal waters may have occurred when there was a temporary rise in the Gulf water level. This is a theory developed by Dr. Carl L. Hubbs of the Scripps Institution of Oceanography and Robert L. Miller of the University of Michigan. Both men have based their explanations on the latest findings.

After the waters of the last high-stage Lake LeConte evaporated, the lowest part of the Salton Sink became but a salt-encrusted playa with only a small salt marsh in its center, kept moist by a few springs and by the occasional runoff of summer cloudbursts and winter-rain freshets. At the turn of the last century the rich alluvial soils of the basin, particularly those of the delta region, were recognized as valuable for agriculture and some of the Colorado River water was diverted onto them for irrigation. To obtain this water, an old dry arroyo was utilized and an opening made in the river bank. No adequate headgate was built—only an intake. For several years the system worked well and real estate boomed in what now became known as Imperial Valley. Later, to get additional much-needed water, a new intake was made below Yuma, in Mexico. That was the beginning of a devastating tragedy that no one at the time could envisage.

In 1905, the river, swollen by flood waters, suddenly began enlarging the intake of the canal system and in a matter of hours great volumes of water were pouring through the growing breach and coursing toward the bottom of Salton Basin. It seemed that nothing could be done to stop the wild inflow of the Colorado. Almost as if by magic, Lake LeConte was being re-created; but now it was to be called the Salton Sea. Dike after dike, each but a makeshift, was hurriedly constructed, only as quickly to be swept away by the raging river. At one time the breach was half a mile wide. Finally, in February 1907, after eighteen months of most desperate herculean struggle, the opening was closed by dumping in brush mats and vast amounts of rock brought in by the Southern Pacific Railroad. At a cost of over $2,000,000 the river had been brought under control and the Imperial Valley saved for future agricultural development.

The present Salton Sea (named after a now-submerged railway siding called Salton) is about 35 miles long and some 15 miles broad at its widest part. Its greatest depth is nearly 47 feet, but much of it is little more than waist deep. Because some old salt deposits on the east side were covered with the rising waters and much of the salt redissolved, and also because the sea has no outlet, it now slightly exceeds ocean water in salinity. Intake from rains and from waste irrigation water brought in by the All-American Canal have recently been exceeding the amount of water lost by evaporation.

At the southern end of the Salton Sea, near Niland, were once most interesting "mud pots" where hot waters, liquid mud, steam, and gases bubbled up constantly or at intervals through crater-like cones of fine silt and sand. All are now submerged by rising water. Similar hot-water vents are located at Volcano Lake on the flood plain of the Colorado delta about 25 miles south of Mexicali, in Baja California. These two areas are thought to be located along the great San Andreas Rift or fault line, which parallels the Salton Sea on the east and extends from the Gulf northward through California to beyond San Francisco. Every few miles along this great earth fissure are springs or seeps, some of cold water and others of scalding-hot water (as at Desert Hot Springs, Hot Mineral Well, and Arrowhead Springs). Many of the desert's natural palm oases occur along this rift, where cool water seeps to the surface.

Between Yuma, Arizona, and the Imperial Valley of California lies a generally north-south series of the most extensive and highest sand dunes to be found on the entire Sonoran Desert. These are the Algodones (Spanish, "cotton") Dunes. They are about 5 miles in width but extend fully 50 miles, from below the Mexican border north along the Southern Pacific Railroad nearly to the town of Niland. The great side-branch of the All-American Canal, bringing irrigation water from the Imperial Dam northward to the lower Coachella Valley farmlands, cuts through, then skirts these magnificent dunes at their western edge. As you travel the highway across the dunes in going from El Centro to Yuma, you are on a wide, well-paved road, but in the early 1900's travelers precariously crossed this long, sandy stretch on a narrow, bumpy plank "roadbed" which had turnouts every quarter of a mile to permit passing. Even today sand-blasted, wrecked portions of this interesting plank road can be seen protruding here and there above the shifting dunes. (Incidentally, it may be mentioned that the Algodones Dunes have been the background for many Hollywood film epics with a Saharan locale.)

The pale yellow sands of the Algodones Dunes are of granitic origin. On their ever-changing surface, often exposed to desiccating heat and nearly daily blasts of wind, it is surprising to find any plants growing at all. However, there are several unique species that not only tenaciously survive but by dint of good luck actually seem to thrive under such adverse conditions. The most conspicuous of these is a sand grass (*Oryzopsis hymenoides*) and a large shrubby wild buckwheat (*Eriogonum deserticola*) which may reach a height of 5 feet and have a stem nearly 2 inches in diameter! This hardy plant has long vertical roots, sometimes over 12 feet long, which penetrate great depths to reach moisture trapped under the sand. Usually the upper portions of the roots, sometimes as much as half of them, are left prominently exposed when the capricious winds move the sand about. When, as sometimes happens, the plant finds it is in danger of being buried by the drifting sands, it slowly elongates its stems to keep above the surface. Certainly the most curious plant of the Algodones Dunes is the strange, rarely seen parasitic "sand food" (*Ammobroma sonorae*), which springs up after unusually wet winters. Its subterranean, succulent stem resembles a swollen felt-covered asparagus stalk, at the top of which appears on the surface of the sand a large thick button-shaped, sand-colored head, the top of which is thickly studded with small purplish flowers opening in successive circles. Roots, stems, and flower heads formerly were much prized as food by the local Cocopa, Papago, and Yuma Indians. The sand food may be seen also in favorable years on the sand hills of northwestern Sonora at the head of the Gulf of California. This plant attacks the roots of near-by host plants such as the sunflower (*Helianthus canus*, 270), plaited leaf (*Coldenia*), croton (*Croton californica*), and the shrubby buckwheat previously mentioned.

Bordering the Imperial Valley far to the westward are the beautiful Laguna Mountains, a part of the Peninsular Range which stretches on southward into Baja California. Hemmed in by these highlands on the west, by the Santa Rosa Mountains on the north and northeast, and by a boundary near the Salton Sea on the east, lies a large desert area of some 272,000 acres which has been set aside by the state of California as Anza-Borrego State Park. It contains several remote native palm oases and brilliantly colored badlands comprising a Californian "painted desert." Borrego State

Park, of 188,760 acres, lies just about north of this. There is a privately owned area in the northern portion of this state park which has become a well-known winter resort. Most of Anza-Borrego State Park, like much of the adjacent desert, consists of a charming desert wasteland of plains and hills covered by ocotillo and creosote bush, and containing the only known natural occurrence in California of the small-leafed elephant tree (*Bursera microphylla,* 223). It includes very rich gypsum deposits and a calcite mine, in addition to thick black-banded beds of fossil marine organisms, especially oysters (*Ostrea lurida*). The Anza Desert State Park was named to commemorate the great Spanish explorer Juan Bautista de Anza, who traversed the western edge of this desert with his followers in 1775 while on their long and arduous overland journey to establish a settlement at San Francisco. Later the famed Butterfield stagecoaches followed this same route into Los Angeles.

Another desert state park has been established at the north end of the Salton Sea. This wonderfully scenic area, called Salton Sea State Park, takes in only a portion of the shore and beach. Adjacent are Box, Painted, and Hidden Springs canyons, where are some of the southwest's finest scenery and vertical-walled gorges. The Mecca Mud Hills in which these gorges are found are not of marine but fresh-water origin. They are composed, for the most part, of stratified non-fossil-bearing layers of Tertiary shales, sandstones, clays, and conglomerates, in many places brilliantly colored and banded, and nearly all tilted and twisted by the action of past uplifting forces. The only fossils known from the beds of the Mecca Mud Hills—other than various bivalve marine shells—are mostly those of mammals, such as horses and camels. Painted Canyon and Box Canyon, the most picturesque of the dozens of side canyons interfingering these hills, are accessible by a road. They are places well worth a visit in the cooler winter months.

At the far northwestern end of the Salton Sink, on the edge of the Coachella Valley, towers lofty San Jacinto Peak (10,832 feet), the north face of which is considered to have the greatest sheer drop of any mountain in the United States. It is the highest peak of the San Jacinto mountain range and forms the southern boundary of the San Gorgonio Pass. North of this pass, in the San Bernardino Mountains, rises massive San Gorgonio Peak, also called Grayback (11,502 feet).

In many of the steep-walled rocky canyons of the Colorado

Desert's bordering ranges and in some of the shallow arroyos of the clay and sandstone hills east and north of the Salton Sea grow small, isolated, and often completely hidden groups of the beautiful robust-trunked desert fan palm (*Washingtonia filifera*). These same palms are also found in canyons along the east base of the San Pedro Mártir and Juárez mountains in adjacent Baja California; there they are often associated with blue palms of the genus *Erythea.* Since the desert fan palm is so largely confined to the Colorado Desert, it is often considered this desert's most distinctive indigenous large plant. The largest and most accessible of the fan palm oases are ensconced in the steep-walled granitic gorges on the east face of the San Jacinto Mountains and in the contorted canyons of metamorphic rock found in the contiguous Santa Rosa Range of California. Of these, Palm Canyon and Andreas Canyon, about 6 miles south of the city of Palm Springs, are reached by good roads. In Palm Canyon it is estimated that about 4,000 wild palms are growing, many of the trees probably well over a hundred years old. A few sizable palm groups occur on the north side of the Coachella Valley in the Indio Mud Hills; those at Thousand Palms and Willis Palms can be easily visited. Other well-known palm groups are in the Borrego State Park area southwest of the Salton Sea. Fire is the great enemy of a palm oasis, and although the flames will not actually kill the trees, they will forever remove the beautiful fawn-colored "leaf skirts" that normally clothe the palm trunk from head to foot. In groups of palms untouched by fire the leaf-skirted trunks form dense "jungles," and it is always a pleasure to crawl or walk in and out among them.

At Palm Springs is located the Palm Springs Desert Museum, devoted exclusively to the interpretation of desert surroundings. Its programs and other educational offerings include field trips, illustrated lectures, motion picture showings, and exhibits of living animals. The displays in the main gallery are frequently changed, so a visit at any time to this active institution is a stimulating experience.

Between Indio in the Coachella Valley and Blythe on the California side of the Colorado River, a highway passes eastward through a very scenic intermountain trough. Some 24 miles east of Indio a side road leads northward through Cottonwood Springs, the Pinto Basin, and other parts of the Joshua Tree National Monument. To the left of the main highway are a series of picturesque ranges: the colorful Eagle Mountains where large deposits of iron ore are mined for the Kaiser Steel Mills at Fontana, the serrated Coxcomb Mountains, and the sharply rising Palen and McCoy mountains. To the right of this road are the very rocky and photogenic Orocopia and Chuckawalla mountains.

In both the Coachella and Imperial valleys of the Colorado Desert, cotton, winter vegetables, grapes, and citrus fruits are extensively cultivated. One of the most important crops, however, is the date, and thousands of acres are devoted to its culture. One may see well-kept date gardens and may visit the packing houses and attractively kept roadside markets where many varieties of choice dates can be purchased. It may be said parenthetically that the date palm (*Phoenix dactilifera*) is not native, but was introduced from North Africa and Arabia at the turn of the century. Here it has thrived as well as in its original habitat.

From the village of Palm Desert extends a highly scenic road called the Palms-to-Pines Highway. It passes westward up into and across the pinyon-clad Santa Rosa Mountains to the high pine forests of the San Jacinto Range. A trip over this gently winding highway gives the traveler an opportunity to study the gradual zonation of plants from desert lowland (Lower Sonoran Life Zone) through the piñon-juniper country (Upper Sonoran), to the zone of Ponderosa yellow pines (Transition). A short climb from Idyllwild, a resort village among the pines, will lead you on up into the even higher Boreal Life Zone. There are very few places so easily accessible where all of these life zones can be seen in such a short lateral distance. The trip is especially rewarding because of the spectacular panoramic view it affords of much of the Salton Sink. There too is always the chance of glimpsing a wild desert bighorn sheep (*Ovis cremnobates*, 134), several small but rigidly protected bands of which live in a state game refuge traversed by this road.

South of the Chuckawalla Mountains is a large triangular area bounded on the southwest by the Chocolate Mountains and on the east by the Colorado River. Its broad plains are drained by numerous branch washes of the Milpitas Wash, also called the Arroyo Seca (Spanish, *seca*, "dry"). Prior to the occupation of this beautiful area by the United States Navy in 1941, both the large-eared burro deer (*Odocoileus hemionus eremicus*) and a few pronghorn antelope (*Antilocapra americana*) were known to roam there.

The only known native specimens of the sahuaro cactus in California are found in a few of the low hills at the far eastern end of the Little Chuckawalla Mountains, also near the potholes along the Colorado River, and on the east side of the Whipple Mountains a short distance above Parker Dam.

A little-used road on the north side of the Coachella Valley, called the Aqueduct or Dillon Road, parallels the great San Andreas Rift or "earthquake fault." Near it are several thermal springs and hot wells. The road was originally made during the construction of the Metropolitan Aqueduct which brings drinking water through open canals, tunnels, and conduits from the Colorado River at Parker Dam to Los Angeles and several other southern California areas.

The northeastern part of Baja California, along the Gulf of California as far south as Bahía de Los Angeles, lies within the rain shadow of the Sierra Juárez and the Sierra San Pedro Mártir, and is but a southern extension of the Colorado Desert of California. This Mexican Colorado Desert can best be studied by traveling from Mexicali south 125 miles to the small village of San Felipe, noted for its shrimp and big-game fishing. Along the first 50 miles of road is a low-lying delta area of rich agricultural land where Mexican farmers grow cotton, sugar beets, corn, and alfalfa. The quaint but often extremely crude huts and houses of these tillers of the soil are interesting. One is impressed by the wide use made of ocotillo stems in the walls and roofs of the simple dwellings, and in the erection of fences and corrals.

At El Mayor the road passes along the Río Hardy, really backwater from the Colorado River. On the west are the conspicuous Cocopa Mountains, around the southern end of which flood waters and occasionally even tidal waters from the Colorado River have flowed northward into a major below-sea-level lake basin similar to the Salton Sink and, like it, cut off from the Gulf of California by the massive delta of the river. It is called Laguna Salada or Laguna Maquata. Ordinarily this inland lake basin is dry, but records indicate that between 1884 and 1929 a lake occupied the basin at least six times. Great numbers of fish, mostly mullet (*Mugil cephalus*), dwelt in the lake after each filling. Each time the laguna dried up, it is said, coyotes came in numbers to feast upon the carcasses of the dead and dying fish.

After leaving the end of the Cocopa Mountains, the highway passes southward over great barren mud flats whitened with dried crusts and crystals of salt. Near by to the southeast is a dazzling white salt flat (*salina*) some 20 miles long. The highly and richly colored volcanic mountains immediately to the west are the Sierra Pinta ("painted mountains").

The traveler now enters a dry and lonely desert, luxuriant with scattered ironwood and palo verde trees, and great ocotillo-covered plains made golden in spring by the numerous flowers of the low shrubby brittle bush (*Encelia farinosa*) and other composites. Now beginning to make their appearance on sand hummocks are scattered colonies of senita cactus (*Cereus thurberi*), a species sparingly represented in the United States by only about fifty native plants, all in Organ Pipe Cactus National Monument. A peculiar tree with dark green leaves and a short, fleshy gray-barked trunk also appears. This is one of the copals (*Elaphrium macdougalii*), locally known as elephant tree. When not too closely scrutinized, it appears, because of the dark tan or purple-red bark of its numerous side limbs and branchlets, somewhat like a squat-crowned, thickly branched apple tree. This elephant tree loses its leaves in late spring, very rapidly regaining them after summer or winter rains.

The sandy bottomed, dry watercourses abound in small shrubs, and after seasons of good rains there is a luxuriant but ephemeral growth of annual plants often overspread with brilliantly colored flowers. The desert plants grow right down to the flats and bluffs overlooking the Gulf, and here often have somewhat larger leaves and more rankly growing stems than plants growing elsewhere on the near-by desert. Many of the sand dunes along the roadsides are covered with short-statured, half-buried mesquite trees, the curious senita cactus, and a sand burr (*Cenchrus palmeri,* 135), which has very spiny green burrs that change to a purple-brown on drying. The spines of the burrs are almost as sharp and stiff as the thorns of a cactus, and may make walking over the dunes very uncomfortable. In the disturbed soil on the shoulders of the road occur numerous plants of the large white-flowered prickly poppy (*Argemone hispida,* 170), apricot mallow (*Sphaeralcea ambigua,* 282), Spanish needle (*Palafoxia linearis,* 325), several varieties of evening primroses (*Oenothera*), sand verbena (*Abronia villosa,* 287), and long slender-stemmed spiderling (*Boerhaavia sp.,* 289).

The 88-mile stretch of sandy, rock-strewn, lowland coastal desert between El Mayor on the Río Hardy and San Felipe is known as El Desierto de Los Chinos ("Desert of the Chinese"). In August of 1902 an ill-fated party of forty-two Orientals and their guides started on foot what proved to be a long disastrous trek between San Felipe and Mexicali. All but eight perished in agonizing deaths due to heat, thirst, and exhaustion. They were poorly informed, ill shod, and badly prepared in every way for such a hazardous journey. The water holes they sought were never found and most of the ill-starred party fell down to die before the journey was half over.

On the east side of the high Sierra Juárez and the Sierra San Pedro Mártir, which border this desert region on the west, are several large scenic canyons with springs or perennial streams of water. In some of their lower boulder-choked reaches are many unusual shrubs and picturesque groups of desert fan palms (*Washingtonia filifera*) and blue palms (*Erythea armata*). Here, the author has recently discovered two new land snails. In the arid valley along the mountain bases there occur fine specimens of the massive cardon (*Pachycereus pringlei,* 158), the large tree cactus that is found more plentifully farther south on the Vizcaíno Desert. Some of the washes contain fine "forests" of large copals, little-leaf palo verde, and ironwood, offering pleasant camp sites.

Summer travel (May through October) in this part of our desert is to be discouraged. The days then are often exceedingly warm (day temperatures of from 110° F. to 115° F. are frequent), and sometimes when storms are moving up along the Gulf the humidity is uncomfortably high. Such weather makes one miserable indeed, especially if the nights are hot, and even the most spectacular scenery cannot be appreciated. Plan your trip here in the clear, warm days of late autumn, winter, or spring.

The season of rains for the Colorado Desert begins about the last week in December. The incoming clouds are brought in by strong westerly or south westerly winds which frequently kick up sand and dust. However, as a rule the winds soon die down after rain begins to fall. The worst and most persistent winds are those of the late spring season. These generally occur when heavy fogs are prevalent in the cismontane valleys along the Pacific Coast. Summer tropical storms, some of them of considerable intensity, at times move up along the Gulf into the southern California deserts, leaving behind varying amounts of moisture.

The Colorado Desert possesses an unusual fauna of notable variety and specialized form. One of its commonest smaller animals is the widely distributed antelope ground squirrel (*Ammospermophilus leucurus,* 118), often locally called desert chipmunk because it has stripes. These white bands end at the shoulder instead of continuing on to the end of the nose as in true chipmunks. The "antelope" part of its common name is given in reference to the white tail, which is carried curved up over the back when running. It is twitched nervously when the animal is excited. This friendly little "ammo" is one of the few American ground squirrels that remain active all winter.

Other mammals include several species of wood or pack rats (*Neotoma,* 119), the bright-eyed small round-tailed ground squirrel (*Citellus tereticaudus,* 117), dainty pocket mice of several kinds (*Perognathus,* 122), nimble white-footed mice (*Peromyscus,* 121), the spotted skunk (*Spilogale arizonae,* 133), the fleet-footed jack rabbit or desert hare (*Lepus californicus deserticola,* 114), the desert coyote (*Canis ochropus estor*), the bobcat (*Lynx eremicus,* 128), the small big-eared desert kit fox (*Vulpes macrotis,* 125), the ringtail "cat" or cacomistle (*Bassariscus astutus,* 124), and many kinds of bats.

Representative birds of this desert include the roadrunner (*Geococcyx californianus,* 79), LeConte thrasher (*Toxostoma lecontei,* 89), Salton Sea song sparrow (*Melospiza melodia saltonis*), phainopepla (*Phainopepla nitens,* 82), Say phoebe (*Sayornis saya,* 63), Gambel or desert quail (*Lophortyx gambeli,* 73), white-rumped shrike or butcher bird (*Lanius excubitoroides,* 85), verdin (*Auriparus flaviceps,* 101), plumbeous gnatcatcher (*Polioptila plumbea,* 88), cactus wren (*Heleodytes brunneicapillus couesi,* 99), and rock wren (*Salpinctes obsoletus,* 102).

The roadrunner is certainly the bird clown of the desert. He may not look it to those who only glimpse him dashing across the highway ahead of their car or running for a moment at breakneck speed along the road before gliding gracefully off into the brush. But if he can be watched about some desert settler's home, where he has become more or less a backyard pet, he will be seen to be a most ludicrous, sportive, and knowing bird. This prankster loves to worry the cat, play with the dog, and get acquainted with all the other farmyard pets. Even in the wild he enjoys jumping about in topsy-turvy manner or running in circles just for the fun of it. Although he eats wild fruits and berries, most of his food is of animal origin. Cleverly he stalks and pounces upon hapless lizards, small snakes, or large insects such as cicadas or grasshoppers. If hungry enough he may attack and kill a rattlesnake by piercing its brain. The favorite haunts of the roadrunner are brush-covered areas where he can find good feeding grounds and ready shelter from avian foes such as hawks. He is most plentiful in shrubby regions along the Colorado and Gila rivers and in somewhat similar situations in Mexico in Chihuahua, Sonora, and Baja California.

The phainopepla, one of the silky flycatchers, is the only small black bird of the open desert; the male can be readily identified at a distance because of his contrasting color, the white patches on the underside of the wing, and the easily seen blue-jay-like head-crest. He is particularly a bird of the mesquite and cat's-claw thickets where mistletoe infestation is heavy. Often he feeds almost entirely upon mistletoe berries. Because only the thin berry pulp is digested, the seeds pass on through the bird unharmed, and if dropped on a favorable site, as on a young tender-barked twig, they may germinate and start a new plant. There is here shown a most interesting animal-plant relationship. The phainopepla is largely dependent on the mistletoe for food, and the plant in turn relies on the bird for dispersal.

On the Colorado Desert the most commonly seen reptiles include the desert iguana or crested lizard (*Dipsosaurus dorsalis,* 54), gridiron-tailed or zebra-tailed lizard (*Callisaurus draconoides,* 35), leopard lizard (*Crotaphytus wisliseni,* 36), collared lizard (*Crotaphytus collaris baileyi,* 42), ocellated sand lizard (*Uma notata,* 38), desert brown-shouldered lizard (*Uta stansburyana elegans,* 37), flat-tailed horned "toad" or lizard (*Phyrnosoma m'callii,* 44), the chuckawalla (*Sauromalus obesus*), sidewinder or horned rattler (*Crotalus cerastes,* 53), desert diamond-back rattlesnake (*Crotalus atrox*), speckled rattler (*Crotalus mitchelli*), desert or rosy boa (*Lichanura rosefusca,* 51), spotted night snake (*Hypsiglena ochrorhynchus*), leaf-nosed snake (*Phyllorhynchus,* 49), glossy or faded snake (*Arizona elegans,* 50), red racer (*Coluber flagellum*), desert king or milk snake (*Lampropeltis getulus*), and the desert gopher or bull snake (*Pituophis catenifer*). The desert tortoise (*Gopherus agassizi,* 43), is seldom seen in the low portions of the Salton Sink, but may be occasionally found in and to the east of the Mecca Mudhills and in the Orocopia and Chuckawalla mountains. Desert snakes are most active during the warm summer nights, from May through September. They seldom venture abroad during the heat of the day since even ten minutes of full summer sun may be sufficient to kill them. At the approach of winter all snakes, lizards (except the small Stansbury uta, 37), and tortoises go beneath the sand or into rock crevices, or dig special burrows where they hibernate for a brief period.

Of the many desert lizards, certainly one of the most amazing is the almost sand-white gridiron-tailed or zebra-tailed lizard, so named because of the dark scorched-appearing bands on the tail. It is the prize runner of all its kind, darting across the sand like a silver streak, at an almost unbelievable speed. So agile and quick is it that it is hardly possible to run one down. When in flight this lizard, like the desert iguana and collared lizard, actually elevates itself on its hind legs. The long tail is curved up as a balancer and the fore limbs are held close to the body. The lizard now appears for all the world like a miniature bipedal dinosaur, speeding lightly over the ground surface.

The beautifully patterned fringe-footed uma or sand lizard is the most perfectly adjusted of all lizards for living in a sandy habitat. The hind feet have very long slender toes, each of which is equipped with a row of elongated scales that overlap with those of the next toe to form a sort of webbed or paddle foot. When pursued the lizard often dives headlong into the sand and literally swims out of sight within a matter of seconds. To facilitate its easy entry into the sands, nature has provided it with a head flattened somewhat like a shovel. For protection the eyelids have thickened scales about their edges and the nostrils have valves to keep out sand particles.

The waters of the northernmost portion of the Gulf of California, adjacent to the Mexican portion of the Colorado Desert, are considerably cooler than those of the subtropical southern portion and support a fauna which in many respects is quite different from that to the south. The warmer southern waters act as a barrier, in much the same way that a mountain range might, and several invertebrate species trapped and isolated in the northern portion of the Gulf are found nowhere else. Peculiarly some of these, particularly some of the spiny lobsters and fishes, are very similar to or identical with those found in the Pacific Ocean on the opposite side of the peninsula. This is thought to indicate that at one time the waters of the lower Gulf were cooler and made possible free passage between the upper Gulf and the Pacific Ocean.

Very recently fossil remains of about 45 different prehistoric land animals have been found in the Anza-Borrego desert area, south of the Salton Sea. The region, once covered by the waters of the Gulf of California, later (perhaps some 500,000 years ago) became a grass- and forest-covered land supporting a wide assortment of grass-, herb-, and tree-feeding animals such as the rare medium-sized ground sloth (*Megalonyx*), a large horse, a giant 16-foot long-limbed camel, a tapir, two types of deer, antelope, pocket gophers, and land tortoises, remains of all of which have been excavated. Bones of a giant condor-like vulture with wing spread of fully 17 feet were among the exciting discoveries.

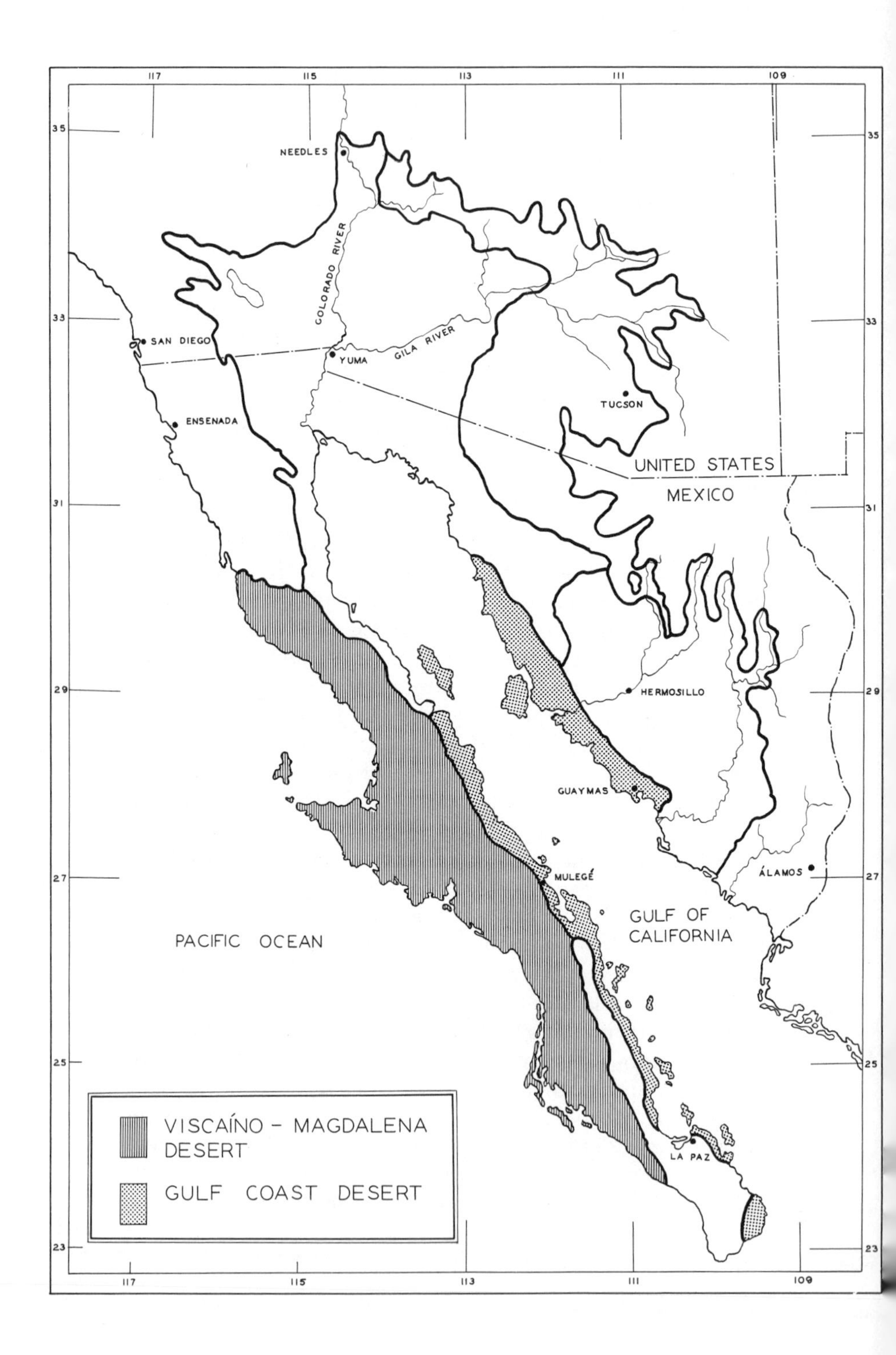

117
115
113
111
109
35
33
31
29
27
25
23
NEEDLES
COLORADO RIVER
SAN DIEGO
YUMA
GILA RIVER
TUCSON
ENSENADA
UNITED STATES
MEXICO
HERMOSILLO
GUAYMAS
MULEGÉ
ÁLAMOS
PACIFIC OCEAN
GULF OF CALIFORNIA
LA PAZ
VISCAÍNO – MAGDALENA DESERT
GULF COAST DESERT

10

THE *Vizcaino-Magdalena* DESERT

The long narrow peninsula of Baja California may be thought of as a much elongated granitic dome or spine, an extension of the great batholith that constitutes the Sierra Nevada of California. This long rugged axis or "backbone" is bordered by metamorphic rocks and covered in many places by lava. It is deeply dissected, particularly around its edges. This more than 700-mile ridged structure is highest at the upper end of the peninsula, where it culminates in the peak called Picacho de la Providencia (10,069 feet) in the pine-clad San Pedro Mártir Mountains. South of these highlands and in the middle third of the peninsula the elevations are much less; near the southern end the broadened ridge ceases, then begins again to form the Sierra de la Laguna or Victoria Mountains with their forests of piñon, oaks, and other broad-crowned trees.

South of the San Pedro Mártir Mountains and a narrow strip of the Mexican Colorado Desert and Gulf Coast Desert lies the marvelously picturesque and interesting VIZCAINO-MAGDALENA DESERT region. It occupies the entire middle section and most of the lower third of the narrow peninsula. Here the desert reaches the very western edge of the continent. Its northern portion was named after the great Spanish explorer, Sebastián Vizcaíno, who in 1596 explored the west coast of Mexico and Baja California; its southern portion received its name from the arid Magdalena Plain which fronts on Magdalena Bay. It is a sparsely populated area of broad and often sandy valleys and outwash plains, scattered boulder-strewn granite hills, and low mountain ridges with steep escarpments. Broad volcanic tablelands, lava flows, and cinder cones, some of very recent origin, here and there break the monotony of the strange landscape. Widely extended plains extend to the east of Scammon Lagoon, the rather low Sierra Vizcaíno, and Ballenas Bay; others almost equally broad front on Magdalena Bay. In parts the soil of this near-level, sun-scorched land is alkaline and exceedingly poor in plant life of any kind. Near Scammon Lagoon and Ballenas Bay are enormous deposits of rock salt. There are large areas where the desolate and forbidding-looking lava is broken into large fragments and is so rough that it is almost impossible to get over it, even on foot.

In the middle and southern reaches of the Vizcaíno-Magdalena region is an elevated, often lava-capped, canyon-dissected mesa country. In parts it is tilted toward the east and underlaid with a horizontal bed of tuffs, sandstone, and other sedimentary rocks. It is known as the Great Volcanic Plateau, and is nearly 300 miles long and in places 30 miles wide. Here groups of cinder cones, extensive lava flows, and bold escarpments give an unreal, lunar appearance to the scene.

The large coastal desert plain, extending north and south and fronting on Magdalena Bay, is over 200 miles long and from 10 to 20 miles wide, and is, for the most part, almost wholly free from hills. It is, next to the Vizcaíno Plains, the largest stretch of almost level land on the peninsula. On the east it is bordered by the low midpeninsula foothills of the interior plateau and farther south by faulted blocks of rock composing the broad mountain belt, known as the Sierra de la Giganta (the culminating peak 5,774 feet), with its tremendous and grandly picturesque scarp face fronting on the Gulf of California. Much of the soil of the Magdalena Plains consists of outwash debris. There are large playas and small mud and clay pans (playitas), some of them filled with water in wet seasons and often afterward thickly covered with a meadow-like growth of grasses, sedges, and colorfully flowered herbaceous perennials.

A little south of the 28th parallel and on the eastern side of the peninsula are the recently active Volcanes de Las Tres Vírgines, three high craters made of black lava, ash, and cinder. Huddled together in a small area, they rise boldly from the surrounding lava plain, the highest being about 6,500 feet. Seen from the gulf side they form a conspicuous and impressive landmark. The last eruption is supposed to have taken place in 1746.

"The broadest and most impressive part of the plateau," to quote Dr. E. W. Nelson, "is located between San Ignacio and Comondú. It is an area of tremendous former volcanic activity. The underlying sandstone is heavily capped by great beds of lava and the surface of the undulating and broken plain is dotted with craters, and over great areas is covered with beds of lava, sometimes in almost unbroken sheets and again in a sea of shattered fragments. Deep canyons cut their ragged way through this dark lava plain, in the two largest and most notable of which lie the oases of La Purísima and Comondú. In these two canyons small streams flow for miles down the sandy and rocky bottoms but elsewhere water is scarce. Vegetation is scanty on the top of the plateau but abundant in the bottoms of all the canyons."

Just north of La Paz a broad sandy desert plain, some 15 miles wide and of less than 100 feet altitude, separates the reddish-brown volcanic Sierra de la Giganta from the pale-gray granitic Sierra Victoria to the south. It is really an eastward extension of the Magdalena Plain and reaches to the shores of La Paz Bay. It is quite evident that in comparatively recent geological time this lowland area was covered by the sea, making an island of the present Cape Region, which is much older geologically than the remainder of the peninsula.

In this whole Vizcaíno-Magdalena region there are but few large springs and these are widely separated. Not one living stream of any volume reaches the sea.

Both winter midlatitude storms and summer rains occur. The summer rains, particularly near the gulf coast and highlands, often come in the form of tremendous downpours, sending temporary torrents coursing down the otherwise dry washes, uprooting plants and carrying unusual amounts of sediments and plant debris. Baja California as a whole has a very meager rainfall, and sometimes for several years in succession no effective drizzling rains may fall over a large portion of the more arid parts of the peninsula. Under these severe conditions there has developed, particularly in this desert, a much specialized and unexpectedly abundant flora which gives to the landscape a strange and often bizarre appearance. The visitor, finding himself suddenly in the midst of so many unfamiliar plant forms, gets an uneasy feeling and can almost imagine himself to have been transferred to another planet. This odd flora is one of the world's richest and most extraordinary assemblages of plants.

The flora of the peninsula is derived from two sources. The plants of the northwest coastal area as far south as El Rosario, the elevated granitic area of the Sierra Juárez, and the Sierra San Pedro Mártir comprise species derived largely from the north. They disappear rapidly to the southward. The southern two-thirds of the peninsula (except the summits of the higher mountains) has a flora which was for the most part derived from the adjacent arid-tropical mainland of Mexico (southern Sonora and Sinaloa).

In traveling southward from the International Boundary to the tip of the peninsula there are three places where one notices a rather abrupt change in the plant cover. The first is at the Río Rosario where one first sees the remarkable idria and cardon. As one passes on south to near Punta Prieta there is a very noticeable enrichment of the vegetation. The traveler sees another decided change near the 28th parallel, just south of El Arco. Here the gradual accession of numerous southern arid-tropical plants such as Palmer's fig (*Ficus palmeri*) and palo blanco (*Lysiloma candida,* 191) is noticed. Frosts are slight and infrequent and there are many plants representing genera of wide distribution farther south. South of the Magdalena Plains is the third divisional line; here one enters the cape thorn forest.

Certainly the most unusual plant of the Vizcaíno portion of this desert is the tree called cirio (*Idria columnaris*); it is restricted to this region and a small area in adjacent Sonora near Puerto Libertad. It is called cirio (Spanish, "candle") by the Mexicans because of its fancied resemblance to the wax candles of their church altars. Here on the peninsula it is one of the most abundant of the larger plants and reaches tree-like proportions. Someone has likened the form of the cirio, especially the young plants, to a long, slender, upside-down carrot or parsnip root branched from the tip.

From the swollen cirio base (sometimes 2 to 3 feet in diameter) rises the gradually tapering pole-like stem, 15 to 40 or even 50 or 60 feet high! This tall trunk has within a "woven" woody tubular framework covered with a smooth, thick, pale yellowish-gray bark. Within is a bitter, soft, and spongy potato-like pulp which is capable of storing large amounts of water. From along the length of this "pole," arranged in rising spirals and protruding from the numerous perforations of the woody framework of the trunk, extend many short, very thorny, pencil-size and often nearly leafless branchlets, each from several inches to 1 or 2 or, rarely, 3 feet in length. On slender stalks at the tree's very top, small yellow tubular flowers sprout forth in early summer. Frequently these peculiar columnar trees branch near their upper third or fourth, giving a pronged or forked appearance; others may be found that are top-heavy and leaning far over, or are grotesquely crooked and malformed because of early injury. Scattered here and there among the living trees are a few dead ones, their bleached white spars standing out in marked contrast to the surrounding green cacti and other shrubs. Sometimes the hollow trunks are inhabited by honey bees. The strangest-appearing of all the cirios are those of the Pacific coastal plain, which have long, dangling gray-green beards of a lacy lichen (*Ramalina reticulata*) draped like Spanish moss on

The elephant tree (*Pachycormus discolor*) likes best the rock areas. Here it grows with cirios and large cacti and agaves of the Viscaíno Desert. On the left *Idria* and young cardons (*Pachycereus pringlei*); center foreground, *Agave nelsoni.*

Dr. Homer Aschmann

Beautiful forests of weird and remarkable idria (*Idria columnaris*) grow on the rocky hills and sandy flats in many places of the Vizcaíno Desert of Baja California. Low volcanic hills in background, granite boulders in foreground. Low clouds and fogs drift in from the Pacific Ocean at night and early morning, supplying the plants with a certain amount of moisture.

A real giant among cardons (*Pachycereus pringlei*). Vizcaíno Desert near the 28th Parallel. Young specimens of *Yucca valida* to the right.

the side branchlets, its growth made possible by the fogs which frequently—especially during June, July, and August—drift in from the Pacific. The pole-like forests of these unique trees cover hundreds of square miles of the interior Vizcaíno plains, hills, and mesas. They are especially impressive when seen growing, as at south of Mármol, among huge granitic boulders. The scene reminds one of quickly burned-over forest land with only the charred dead tree-trunks remaining. In moonlight they look like gray ghosts studding the rocky hillsides.

It may be added parenthetically that the semisucculent cirio is placed together with the ocotillo in the plant family *Fouquieriaceae.* Superficially there is only slight resemblance except in the form of the flower and the origin of the thorns which in both plants are curiously developed from the indurated midribs of the primary leaves of the previous season.

During years of extreme drought when ordinary feed is scarce, the Mexican cattlemen cut down many of the cirios and split them open so that hungry cattle can eat the succulent pith.

Many kinds of cacti abound. They vary in size from the giant cardon (*Pachycereus pringlei*, 158), probably the largest cactus in the world, with massive spiny and fluted trunk that may rise as high as 60 feet, down to the smaller cylindrical-stemmed opuntias of many kinds, flat-leafed opuntias, small nipple cacti, barrel cacti, and some of the weak straggly-stemmed, ground-hugging species. Both the larger and smaller cacti may grow close together, and when intermingled with cirio, ocotillo, and agave sometimes form a thorn jungle impossible to penetrate without an axe or machete. The fruits and handsome flowers of many of these cacti are edible and for this reason are eagerly sought by both birds and small mammals. They were a main article of diet of the Indians who formerly occupied these lonely tracts.

The impressive and always conspicuous giant cardon is a close relative to the sahuaro cactus of Arizona and adjacent Sonora. It differs from the sahuaro in being more massive (some of the plants may weigh up to 10 tons), more compactly branched, and having its center of gravity nearer the ground. The long upsweeping arms as a rule branch off at sharp angles rather than in graceful curves. At the north end of the Vizcaíno-Magdalena Desert entire hillsides are often covered with crowded forests of mixed cardon and cirio. In the southern Magdalena area this cactus is often shorter, sparingly branched, and spaced more widely. The Mearns' gilded flicker and gray breasted cardon woodpecker often dwell about these strange succulent trees, excavating holes in them for their nests.

Speaking of the cardon, the Jesuit missionary, Father Johann Jakob Baegert wrote two hundred years ago: "It grows six fathoms high, and its branches are nothing more than twelve to fifteen green round beams, three to four spans thick with channels or furrows running the full length clear around the evenly shaped stems. They are about as high as four or five men; when young these branches are covered with spines. Their substance is not hard and durable, like wood, but decays in a few weeks when cut off, falls apart and becomes a soft pulp. Nothing is left but a poor, useless skeleton. This terrifying scaffold in the shape of a plant is scaled by the natives either to pick the tasteless fruit which grows out of the sides of those beams or to survey the country in search of game, heedless of the fact that the whole structure shakes and bends from side to side because its substance is so weak."

Only those who penetrate deep into this arid wilderness of thorny vegetation have opportunity to see and fully appreciate the charm of such strange plants as the caterpillar cactus, or devil cactus (*Lemaireocereus eruca*), one of the oddest of cacti. It was named in honor of the French horticulturist, Charles Lemaire, and occurs only on the plains to the east of Magdalena Bay. Here it may be found in "beds" resembling collections of hideous giant spiny worms, each of the radiating stems apparently moving outward from a common center. The parent stems grow prostrate on top of the ground and send down strong rootlets all along their length.

"Its manner of growth with uplifted heads and prominent reflexed spines gives the plant a resemblance to huge caterpillars," wrote T. S. Brandegee, the botanist who first described it. "The resemblance is especially striking when the plant meets with some obstruction such as a log or large stone. Then it raises its head, crawls up one side and down the other, and finally, by the dying of the rear, virtually passes over the obstruction!" As the older back parts of this unbelievably strange cactus die, the front ends of the stems elongate, continuing to form new roots as they "creep" forward, leaving the original dead parts behind. In this way, many living portions come to lie far apart from the parent colony. What a peculiar adaptation for the dispersal of this fantastic cactus!

Other conspicuous cacti are the 3 to 9 foot high, very compactly branched, light-green-flowered cochal (*Myrtillocactus cochal*), with red globular fruits, and the very common and widely distributed cholla (*Opuntia cholla*), with its very spiny, many-branched heavy stems, deep purple flowers, and fruits which hang in short chains. In the central Vizcaíno Desert flourishes the openly branched *Opuntia clavellina*, with slender joints covered with long yellow to brown spines; the flowers are yellow. In the northern Vizcaíno area is found the tall, slender, maroon-flowered barrel cactus, distinguished by its beautiful deep-rose-colored spines.

Certain to evoke interest is the dark-colored cactus called pitahaya agria (*Machaerocereus gummosus*, 352), a species closely related to the organ pipe cactus. It is probably the most widely distributed of all Baja California cacti, everywhere mingling with the thorny brush and trees. Some of the long, stout, dark purplish-green, 8- or 9-ribbed stems lie prostrate. Others strike out in different directions "without order or symmetry," as wrote the Mexican historian Clavigero. The entire plant may have a spread of 18 or 20 feet. Like the caterpillar cactus, with which it is sometimes confused, the prostrate ground-hugging portions of the stem send down numerous anchoring roots. Long purple flowers appear in midsummer, these followed by bright scarlet, purple-meated, slightly tart fruits which are considered by the natives the choicest and most dependable of all local desert cactus fruits. Unfortunately these cactus apples are not abundant as are those of the "pitahaya dulce," known familiarly as organ pipe cactus.

In parts of the Magdalena Desert the pitahaya agria often consorts in such abundance with the thorny, white-barked tropical tree called lysiloma that biologists sometimes refer to this area of Baja California as the Lysiloma-Machaerocereus Region.

Lysiloma candida (191), called "palo blanco" by the Mexicans because of its smooth white bark, is a tree 20 to 25 feet high. As in many other leguminous trees, the leaves are bipinnate with many very small leaflets. The numerous yellow-stemmed flowers form small but rather showy ball-like heads. The fruit is a flat bean 3 to 6 inches long. Natives use the bark for tanning and formerly exported much of it. Leathers tanned with lysiloma bark are of unusual excellence and much sought by exporters.

Another strange woody plant is the torote or elephant tree (*Pachycormus discolor*), not to be confused with the little-leaf copal (*Bursera microphylla*) or the gray-barked copal (*Elaphrium macdougalii*, 292), both of which are also locally called elephant tree. They are common in certain parts of the peninsula, and also in parts of western Sonora on the mainland of Mexico. Like the cirio, the torote is largely confined to the rocky slopes and plains of the midpeninsula, from San Fernando southward. Off the west coast, in the Pacific, it also occurs on the Magdalena Islands and on Santa Margarita and Cedros islands. The short, puffy, fat-stemmed trunk, covered with smooth, thin, light buff-colored bark, may be up to 2 feet in diameter. From this swollen basal trunk extend thick, wide-spreading, tortuous limbs of exaggerated thickness, tapering abruptly into the short leaf-and-flower-bearing branches. It is these thick bulging limbs that suggest an elephant's trunk or its massive legs.

The dainty panicles of bright pink or sometimes yellow flowers appear after the shedding of the bright green leaves. When the rarely blossoming trees are grouped in masses they form a blaze of color visible for several miles. The bark of both the main trunk and the branches scales off as the tree grows in size. The compound leaves, which appear shortly after heavy showers, soon drop off, so that for most of the year the elephant tree appears quite lifeless. This is an adaptation to conserve every bit of possible moisture. Several other desert plants, including the ocotillo, show this same adjustment to environmental conditions. In coastal areas subject to fog the trees often carry dozens of gray balls of epiphytic ball moss (*Tillandsia recurvata*) or are partially "smothered" by the parasitic love vine (*Cuscuta veatchia*).

Along the courses of many of the steep-walled ravines or barrancas of the mountains and plateaus, where water comes very nearly or directly to the surface, as at San Fernando and Catavina, graceful fan palms (*Washingtonia* and *Erythea*) grow scattered and in colonies, from small stands to colonies of several hundred trees. Often the palms are found about water seeps in combination with such strange plant neighbors as the tall cirio and thorny ocotillo, the whole assemblage forming a scene at once beautiful and appealing, as well as seemingly incongruous.

On the steep rocky walls of the deep-cut barrancas of the volcanic mesas, both of the southern part of the Gulf Coast Desert and the less arid areas of the Magdalena region, one may find the white-trunked wild fig or zalate (*Ficus palmeri*), a tree of most peculiar growth. It extends its lower smooth-barked, eel-like, much-flattened stems in fantastic manner over the surface of the naked rocks and into the clefts where small rootlets penetrate to seek moisture. Unlike the fig of commerce, its leaves are conspicuously longer than they are broad; the small pear-shaped figs are almost flavorless but nevertheless sometimes eaten. Trees growing in the good soil of canyon bottoms are often quite large and well shaped.

The handsome peninsular tree yucca (*Yucca valida*), called datilla, is a near relative of the Joshua tree of the Mohave Desert and much resembles that well-known yucca in general form. It flourishes from the northern parts of this midpeninsula desert to the Cape Region, growing in greatest abundance in the deep-soiled valleys of the Vizcaíno plains, where it may form extensive forests. It often constitutes an impressive feature of the landscape, especially when growing in thickset stands or "jungles" with the cirio and long-stemmed organ pipe cactus, as well as the pitahaya agria, ocotillo, and agave (*Agave shawii,* or *Agave nelsoni,* 196). The taller tree yuccas of the coastal area do not long remain vertical, but because of the prevailing westerly winds start to lean after attaining a height of but 10 or 12 feet.

Among the abundant, curious, and widespread shrubs of this desert, and also of western Sonora, is the low- to medium-sized evergreen, upright, malodorous guayacán, the *Viscainoa geniculata* (242) of the botanists. It often grows in thickets, along the edge of dry streamways in association with cirios, mesquites, and goat-nut. The rigid branches are stout and crooked and bear broad 5-petaled, yellowish-white flowers, followed by conspicuous big, usually 4-lobed, capsular fruits. The rather large ovoid ashy-green and leather-textured alternate leaves are usually simple, but pinnate leaves with 3 to 5 leaflets are sometimes found. It is a close relative of the familiar creosote bush, and like that shrub belongs to the lignum-vitae family.

In the middle and southern part of the peninsula, as well as in parts of Sonora and Sinaloa near the coast, is a peculiar type of ocotillo often found in sandy plains. It differs from the ocotillo so often seen in the deserts of the southwestern United States and far northern Sonora, by having regularly a very short thick trunk and only a few widely divergent stout main stems, many small slender crooked end-branchlets and branches, and a redder flower. This spiny shrub, called the peninsular ocotillo (*Fouquieria peninsularis*), was mentioned first by the Mexican historian, Francisco Clavigero, whose *Storia della California* was published in 1789: "There is another tree bristling with spines, and almost always naked, for which reason the Spaniards gave it the name of palo Adán ('Adam's tree'). When there is rain it sends forth a few small leaves, but after a month it sheds them and remains naked all the year." Near the west coast the thorny stems are almost completely covered with small tufted gray-green lichens of the genus *Ramalina.*

The mammals, birds, and reptiles of the Vizcaíno-Magdalena Desert include many genera common to the southwestern deserts of the United States. However, due to their long isolation, many of them show sufficient differences to require their classification in different species and very often in different subspecies or varieties, probably distinguishable only to the trained biologist. Several genera are absent and a number of new ones are present, these last represented by forms which have migrated northward from the Cape District. Kit foxes, coyotes, jack rabbits, flat-headed wood rats, and California valley quail and numerous other birds are likely to be frequent camp visitors, especially if the traveler stops near springs or small streams. Gray foxes, ring-tailed cats, and striped skunks come in to feed at night on dates when date palms are found about such oases as occur at Mulegé and San Ignacio.

The thorny agave thickets, which often extend over vast areas, are the habitations of numerous rodents, especially wild mice. There these resourceful creatures find not only adequate shelter but plenty of food and, in the succulent leaves, readily available water.

The seeming abundance of larger game animals is false. The number per square mile is really very small and constantly dwindling. It is only the lack of widespread hunting that makes these animals as evident as they are. Already meat-hunting miners and trophy-hunting sportsmen have almost wiped out the antelope and soon the deer and desert bighorn will go the same way. It is interesting to note that the first desert bighorn discovered in America were those recorded in the writings of the Spanish missionaries working in Baja California.

Two species of horned lizards (*Phrynosoma platyrhinos* and *P. solare*) are found, as well as the swift-running gridiron-tailed lizard so common on desert plains to the north.

It has been estimated that originally there were probably about 40,000 Indians living in Baja California, an average of nearly one per square mile. Most of the tribes to which they belonged are now extinct, and little evidence remains of their culture, simple as it must have been. Inhabiting mainly the coastal desertland were such little-known tribes as the Kiliwi, Cochimi, Waicuri, and Pericu. A few Cocopa, members of the only tribe in Baja California to practice agriculture, still live about the delta of the

Colorado River. Our records of the miserable form of life of the Baja California Indians is to be found in the often very detailed and picturesque reports made periodically by the Jesuit, Franciscan, and Dominican fathers who explored the entire length of the peninsula and established missions along most of its extent. Most of these missions are now in total ruin. Of the remaining ones, those at San Xavier, San Ignacio, San Borja, and Mulegé are best preserved. The mission at Loreto has been extensively remodeled and practically rebuilt since the earthquake of 1877.

The Jesuits, led by Fray Juan Salvatierra, established the first Baja California settlement on the Gulf in 1697 at Loreto, which was to be the ecclesiastical and political capital for more than a hundred years. They found the Indians a miserable, almost completely savage people, unclothed and without knowledge of the most elementary principles of agriculture. These aborigines they gathered around the missions which they established about the chief watering places. They instructed them in the Catholic faith, and taught them to do masonry and carpenter work, and to tend cattle and cultivate the fields. Mission buildings and also pueblos were constructed with Indian aid. Indians also built a network of well-made foot trails connecting these pioneer establishments. Some of the trails can yet be followed, especially where made over places where lava blocks had to be lifted and thrown to one side to open the way. There were three main trails extending almost the length of the peninsula; one along each coast and the third in the mountainous interior.

A summary of the achievements of the Jesuits is thus given by Arthur Wallbridge North in his book, *The Mother of California*:

"During their seventy years sojourn in Lower (or Baja) California, the Jesuits had charted the east coast and explored the east and west coasts of the Peninsula and the islands adjacent thereto; they had explored the interior to the thirty-first parallel of north latitude in a manner that has never been excelled . . . they had founded twenty-three mission establishments, of which fourteen had proven successful; they had erected structures of stone and beautified them; they had formulated a system of mission life never thereafter surpassed; they had not only instructed the Indians in religious matters, but had taught them many of the useful arts; they had made a network of open trails connecting the missions with each other and with Loreto; they had taken scientific and geographical notes concerning the country and prepared ethnological reports on the native races; they had cultivated and planted the arable acres and inaugurated a system of irrigation. . . . Of their labor in the Peninsula, it has been said with truth that remote as was the land and small the nation, there are few chapters in the history of the world on which the mind can turn with so sincere an admiration."

In 1767 the Jesuits were ordered to be expelled from all the realms of Spain. However, it was not until 1768 that the expulsion was effected in remote Baja California. Five months later their native charges and establishments were given over to the friars of the Franciscan order, headed by Junípero Serra, who founded the mission of San Fernando. The work of the Franciscans was short-lived (1768–73); then the Dominicans took over and founded nine more missions, all of which are now in ruins; the Dominicans must be given credit for having rebuilt of permanent stone some of the adobe missions put up by the Jesuits. Epidemics of disease in the ensuing years so reduced the Indian population that few were left at the time of the secularization of the church establishment in 1833. By the middle of the nineteenth century the southern Indians had wholly disappeared and only about a thousand remained in the northern peninsula.

The exceedingly dry middle third of the peninsula was able to carry a larger Indian population than might be expected, because there is here an overlapping of the often erratic winter and summer rainy seasons, occasionally making possible a crop of seed-yielding wild flowers twice a year. The prevalence of drought-resisting cacti with their abundant summer fruits, and agaves of many species, the whole young plants of which were sometimes roasted, was an added factor which made living possible and less precarious in this ever-thirsty land. Both cacti and agaves are largely unaffected by the vagaries of the climate.

Pottery making among the Baja California Indians was largely confined to the far northern areas. Virtually no pottery, not even broken shards, so often found in southern California desert areas, is found below the 30th parallel. This lack is probably related to the incessant nomadism of Indians who roamed the middle and southern parts of the peninsula.

At present there is but a single main road down the full length of Baja California, and it is paved only from Tijuana to near San Telmo on the Pacific side. This is followed by some 30 miles of graded and graveled roadbed. After this the road becomes very poor indeed, some of it impassible during rainy weather. Deep ruts, high rocky centers, steep rough pitches, narrow rocky grades, soft spots, thick stifling dust, heavy sand, loose volcanic ash, and lava-strewn plains alternate with fairly good stretches from the end of the graveled road across the peninsula to the Gulf at Santa Rosalía. From Santa Rosalía it is a little better going but much of the road must still be considered very much as an "obstacle course." After leaving Comondú the road passes through monotonous dusty-gray, shrub-dotted sands or over dry lakes of the Magdalena plains. Gasoline, oil, good drinking water, and other supplies are to be had only at long intervals. Enticing as the journey may sound to the desert traveler, only those who carry adequate supplies and have trucks or station wagons with high clearance and equipped with four-wheel drive and good heavy-duty tires should attempt the trip. Small jeeps with their narrow wheel bases can always negotiate the high-centered roads, because they do not slip so deeply into the standard-width ruts but use one side of the high center and thus avoid striking projecting rocks which menace the crankcase and fuel tank of ordinary cars; but they have the drawback of very limited space for carrying supplies.

The road is poorly maintained and is now used primarily by truckers carrying mining supplies, ore, onyx slabs, and cattle, and secondarily by those adventuresome Americans (naturalists, prospectors, and sightseers) who are willing to undergo hardship, slow travel (usually only a few miles a day), and primitive living. Excellent camps amid beautiful and unusual desert surroundings are abundant. The map issued by the Automobile Club of Southern California shows not only the main roads but also the places where supplies may usually be purchased. Gerhard and Gulick's *Lower California Guidebook* is helpful, with its maps, itineraries, and notes on places of historical interest.

The best seasons for travel in mid-desert parts of Baja California are the autumn, spring, and early summer. Because of the prevailing westerly sea breezes, the days of May and June are seldom excessively warm, and the nights are almost always pleasant. In this middle section of the peninsula the time of general profuse flowering of plants sometimes follows the winter rains of December and January; at other times, the heavy tropical rains of July and August. In May and June the cacti, the tree yuccas, and also many of the shrubs are often in bloom.

11

THE *Gulf Coast* DESERT

A very long and narrow strip on either side of the Gulf of California comprises a distinctive arid region, which may be called the GULF COAST DESERT. The part on the east or Mexican mainland side of the Gulf lies adjacent to the southern sandy stretches of the barren Yuman Desert and the near-by plains of Sonora. The part lying along the west coast of the Gulf reaches with slight interruptions from the 28th parallel as far south as San José del Cabo near the tip of the peninsula. (See map, page 104.)

The whole area is in the rain shadow of the peninsular uplands and the winter storms give it little rain. The winter rainfall of the Sonoran central gulf coast is about the same as that of the west side of the Gulf. Here it is the summer rains which are important. At Guaymas there may be as much as three inches in August alone. These summer rains are sporadic and come as northern extensions of storms spreading from the mountainous region of the Cape District. An alternate name for this region is given by Dr. Forrest Shreve, who designates it the Central Gulf Desert and considers it the driest portion of the Sonoran Desert.

The northern part of this Gulf-side desert, on the Sonoran side, consists of low sandy or gravelly plains and of bajadas extending far inland. Southward from a point 50 miles above Guaymas, low mountains occur, rising directly from the shore. To the east of Guaymas are large salt flats, just inland from which is a notable stand of the massive cardon (*Pachycereus pringlei*), the trees made especially conspicuous by the thick white coatings of excrement left by the countless sea birds which habitually roost on their tops.

A narrow tongue of arid-tropical vegetation extends northward through Guaymas to a point north and opposite Tiburón Island. The vicinity of Guaymas is a point of special interest to biologists because it appears to be the meeting point where a number of northern and southern races of birds and mammals come together.

The area was much frequented by both plant and animal collectors in the late nineteenth century.

Near Puerto Libertad is a small isolated colony of some 2,000 trees of the cirio (*Idria columnaris*—see page 108), marking the only occurrence of this singular plant outside the Vizcaíno region of Baja California.

The narrow middle and southern portion of the Gulf Coast Desert on the Baja California side is little known and is at present difficult of access. The plant covering is for the most part sparse and in general is similar in character to that of the opposite coast of Sonora.

The desert ironwood (*Olneya tesota,* 355), so frequently seen in southern Arizona, Sonora, and southeastern California, is widely distributed all along the hot Gulf coast, even as far south as Cape San Lucas at the southern tip of the peninsula.

The indigo-flowered and very spiny smoke tree (*Dalea spinosa,* 348), so plentiful in the low sandy washes of the Colorado Desert, grows in the dry sandy drainage channels or arroyos along the Gulf coast as far south as the middle of the peninsula. The little-leaf elephant tree (*Bursera*) and jatropha are also present.

The direct influence of the humidifying waters of the Gulf of California extends inland only a short distance. Along the Sonoran coast, within reach of the salt spray, grows a small-leaved gray shrub known to botanists as *Frankenia palmeri.* With it may be found the common gray-leaved salt bush (*Atriplex canescens*). The two species form an almost continuous marginal belt which gives a monotonous aspect to the lonely shore.

A long series of uninhabited desert islands border the peninsular coast in the Gulf of California. Almost without exception they are high and rocky, with steep-pitched cliffs along the shoreline. A few consist of granitic rocks but most are of volcanic origin and composed of basalt, ash, and related rocks. Few have any fresh water and on most very severe desert conditions continually prevail. These lonely islands have probably been separated from the peninsular mainland for a long time, and it appears that their plants and animals, with the exception of birds, must have been brought to them by the winds and waves accompanying the fierce storms which from time to time sweep over the Gulf.

The Gulf waters are much warmer than those of the open ocean farther south, and in them flourish an abundance of unique marine life, especially of mollusks and fishes, which in turn encourages a rich representation of sea birds. In great flocks the birds forage along the shores and tidal flats, to share with coyotes the plentiful food supply carried up on the beach with each high tide. It is a strange combination, this meeting place of the seemingly austere and lifeless desert and the turbulent sea with its abundance of teeming life.

Wherever permanent springs are found near the beach very noticeable and very old Indian shell-heaps occur. Some are several hundreds of yards long, as much as thirty feet high and perhaps a hundred feet across. One of the largest of these is at San Lucas south of Santa Rosalía. It clearly demonstrates how much the aboriginal desert-dwelling people depended upon the sea as a source of food.

12

THE *Mohave* DESERT

Wedged in between the Sonoran Desert on the south and the Great Basin Sagebrush Desert on the north lies a more or less quadrangular and comparatively small, shrub-dominated area called the MOHAVE DESERT (sometimes spelled Mojave). Mohave is a name originally given to a tribe of Indians living along the Colorado River on the far eastern margin of this desert. Later the name was used by Captain John C. Frémont to designate a river which traverses the area, and which for a long time was thought to empty into the Colorado River in Mohave Indian territory. Actually it empties into Soda Lake.

The cresote bush (*Larrea divaricata,* 238), growing widely spaced and in almost pure extensive stands or often mixed with the burro bush (*Franseria dumosa,* 251) and other low shrubs, here reaches the northern limit of its widespread distribution. The Mohave Desert is the special domain of the curious Joshua tree or tree yucca (*Yucca brevifolia*), and its eastern, slender-trunked, short-needled form (*Y. brevifolia jaegeriana*), a large grotesquely branched, spine-studded yucca found nowhere else. It is indeed the Mohave Desert's most distinctive plant. If a line is drawn around the outer limits of this strange tree's distribution, that line pretty well marks out the marginal confines of the Mohave Desert region.

This is an upland desert with elevations for the most part between 2,000 and 5,000 feet above sea level, but in one part of it, Death Valley, approximately 550 square miles lie below sea level. Near Tule Spring and west of Natural Bridge Canyon in Death Valley are found the lowest spots in the western hemisphere, 282 feet below the surface of the sea. Lying within the rain shadow of the southern Sierra Nevada, the Tehachapi, the San Gabriel, the San Bernardino, and the Little San Bernardino mountains, the Mohave Desert receives but a meager rainfall which measures between 1.4 and 5 inches per year. Most of this moisture falls in winter and spring. During the colder months snow may blanket

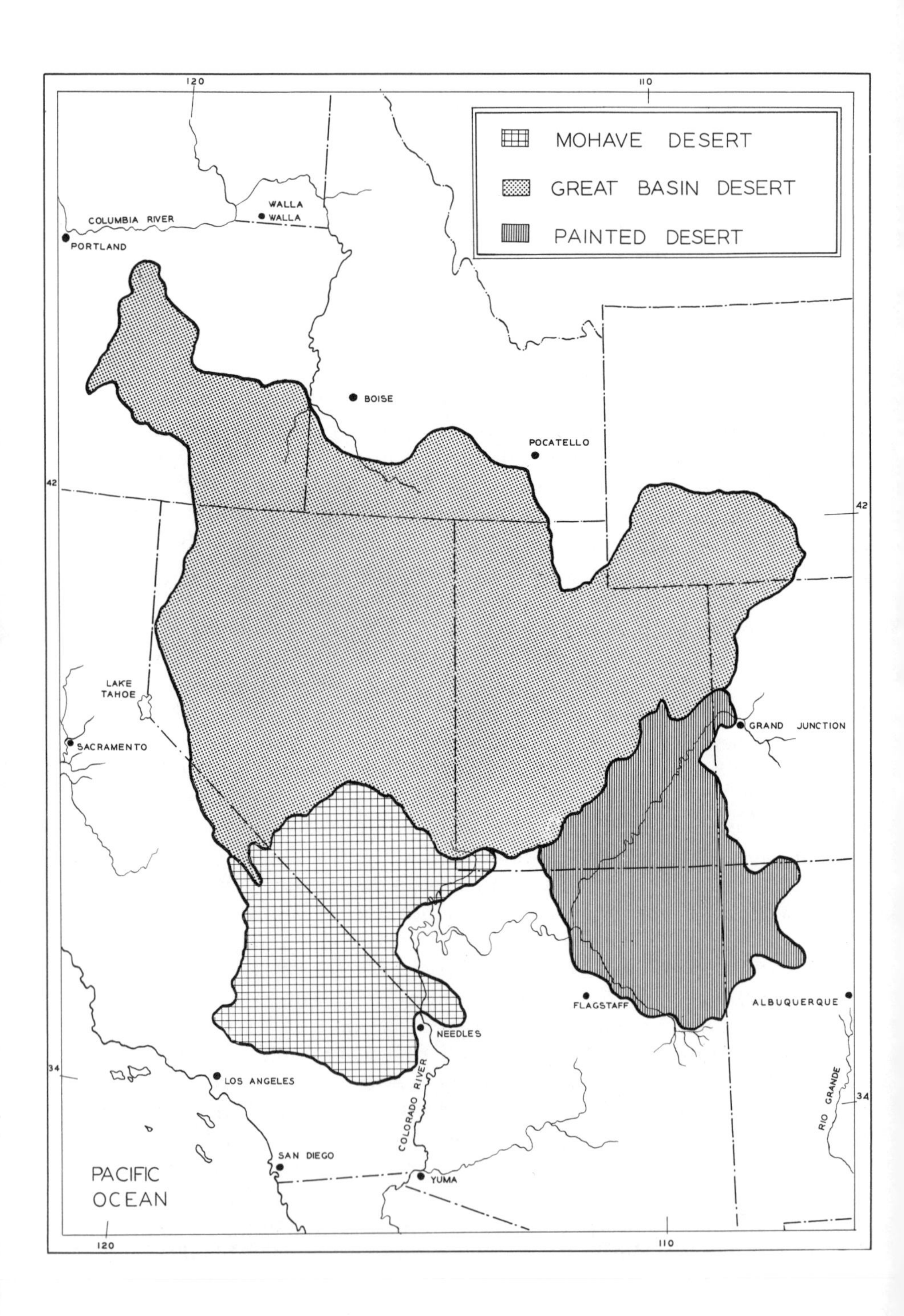

MOHAVE DESERT
GREAT BASIN DESERT
PAINTED DESERT
120
110
42
34
COLUMBIA RIVER
WALLA WALLA
PORTLAND
BOISE
POCATELLO
LAKE TAHOE
SACRAMENTO
GRAND JUNCTION
NEEDLES
FLAGSTAFF
ALBUQUERQUE
LOS ANGELES
COLORADO RIVER
RIO GRANDE
SAN DIEGO
YUMA
PACIFIC OCEAN

parts of the area a number of times, but the melting is rapid. Summer cloudbursts of great intensity and with great erosional power occur somewhere within the limits of this desert almost every year. The storm waters rush down the canyons, carrying quantities of plant debris, rocks, and sand, washing out roads and trails, and often considerably altering the face of the land within a matter of hours. The places visited by these summer storms is later marked by localized areas of bright green creosote bushes, especially noticeable on the pediments of rock, sand, and gravel fanning out from the mouths of canyons which carry the runoff waters.

The Mohave's western and geologically older portion has a surface that is comparatively smooth; the middle and eastern sections are dominated by numerous, more recently formed parallel ranges of sharp angular-sloped mountains with extensive, gently sloping bajadas. The directional trend of most of these barren ranges, and of the basins between them, is from north to south.

Physiographically most of the Mohave Desert is a part of the Great Basin Province. In the long-ago past, at the end of the Ice Age, the more southern and warmer lands, not covered by glaciers, received heavy rains, at times almost of deluge proportions. This time of generous precipitation has been called the Pluvial Period. The Mohave, like much of Nevada and Utah, was then a land of many lakes and rivers fed by the abundant rains and the waters derived from melting snow and glacial ice in the higher mountains about its perimeter. The lowest of the pluvial lakes was Lake Manly in Death Valley, which served as the great sink into which at least three desert streams drained: the Mohave River, arising in the San Bernardino Mountains; the Amargosa River, flowing from the low ranges of southwestern Nevada; and the Owens River, coming from the eastern slope of the Sierra Nevada. These rivers, together with the several large lakes which once impounded water along their courses, were at one time united into an integrated drainage system which endured from postglacial times to within recent years, perhaps even as late as a few centuries ago.

Evidence of this stream-and-lake integration can be seen in the recent fish fauna found in many of the isolated springs and in the few small streams now flowing over the old, now mostly dry, beds of the rivers of the ancient Death Valley drainage system. These fish are all small minnow-like forms (*Cyprinodon*) and are, except in a few instances, closely related. Although they are now isolated from one another by many miles of desert, their presence indicates that there must have at one time been free passage between parts of a continuous stream system such as we have envisaged in connection with Lake Manly in Death Valley.

Water still flows in portions of the old Mohavean river beds, but only the saline Amargosa (Spanish, "bitter"), once appropriately described as "an artery of salt running through the desert," now discharges into the Death Valley sink, and then only during times of heavy flow resulting from flash floods or abundant winter rains. The ancient lake in Death Valley now generally referred to as Lake Manly (in commemoration of William Lewis Manly, who in 1849 led a party of immigrants out of that dread land of thirst) was fully 90 miles long and 600 feet deep. It is possible that Lake Manly drained southwest to the Colorado River. It was probably the first of the large Mohavean pluvial lakes to dry up. The lakes which were part of the Mohave river-lake system occupied the present sites of Panamint, Searles, Silver, Soda, China, and Manix Lake basins. As the climate became more arid they gradually dwindled in size, and today all we see in their places are salt- and alkali-encrusted playas or flat clay beds. Above on the mountain sides, at least in places, are found the old shore lines, wave-cut terraces, and ancient beaches which register the heights once reached by their waters.

It is indeed difficult for the modern traveler, crossing the Mohave's broad arid wastes over a paved highway, to picture this region as once being a land of meandering rivers and numerous lakes supporting a varied and abundant flora and fauna. Fossils found in caves and in deposits of volcanic ash indicate that large mammals such as the giant ground sloth (*Nothrotherium*), the mammoth, the camel, and even the three-toed horse once lived there in great herds, and that one of them at least, the ground sloth, actually existed contemporaneously with the primitive Pinto Basin Man who wandered about this region some 9,000 years ago! Artifacts made by this primitive people found in the Pinto Basin of the Joshua Tree National Monument (hence the name Pinto Culture), and more extensively at Little Lake on the northeastern Mohave Desert, indicate that they were nomadic hunters and so primitive that they did not even possess the bow and arrow; they hunted with the "atlatl" spear-thrower instead. It is not certain that they knew the art of pottery; no shards have been found.

In the many well-formed cinder cones, in the little-weathered and extensive lava flows, and in the layered deposits of volcanic ash, we see ample evidence of widespread volcanism in the Mohave Desert's not too distant past. Symmetrical craters, cones, and congealed streams of black lava are especially prominent near Little Lake at the lower end of Owens Valley, as well as near Amboy. Just northwest of Kelso there are twenty-two beautiful symmetrical cinder cones, all huddled together in a small area.

In the northeastern Mohave Desert, near the California-Nevada border and in the vicinity of Las Vegas in Nevada, are marvelously picturesque mountains of gray, almost solid limestone, many of them beautifully banded with strata lying at every angle between horizontal and vertical. In some of these mountains there are subterranean caverns with stalactites and stalagmites. Since many of the strata have been tilted on edge considerably away from horizontal, the caverns have developed on the form of a series of somewhat narrow rooms, more or less one above another. The Spring (also called Charleston) Mountains, Clark Mountain, and Sheep Mountain, all formed of limestone, are of sufficient height to favor the growth of forests of coniferous trees. From their summits one can obtain magnificent views of the beautiful and delicately tinted arid country surrounding them.

In extreme southwestern Utah, and near Lake Mead and at the southern end of the Spring Mountains, in Nevada, are picturesque layered beds of bright-red sandstone, really portions of an exposed ancient desert whose sands were cemented together by minerals from a sea which once covered them. The Valley of Fire, in Nevada, is probably the most spectacular of these formations. At Overton, Nevada, is a museum where are exhibited many Indian artifacts and other aboriginal remains which have been unearthed from the site of a large near-by prehistoric village.

Near Kelso, California, are found some of the largest, highest, and most extensive sand dunes on our American deserts. The high hills of sand are composed of tawny white granitic particles. Like most dunes they are best viewed in the early morning or evening, or by moonlight. In spring, following a wet winter, their lower parts are carpeted with colorful wild flowers including sand verbenas (*Abronia*, 287), blazing stars (*Mentzelia*, 211, 213), bush penstemon (230), and both yellow- and white-flowered evening primroses (*Oenothera*).

There is only one large river flowing through the Mohave Desert, the Colorado. Several huge dams have been built along its lower course to impound the waters as a means of flood control, to supply irrigation and drinking water, and to furnish energy for the generation of electrical power. Visitors to the Mohave Desert should not fail to take a boat ride on either Lake Mead or Lake

Havasu, the artificial lakes created by Hoover and Parker dams. The gorgeous scenery along the colorful cliffs and the reflections in the blue lake-waters are never-to-be-forgotten sights.

For most visitors, the chief attraction of the Mohave Desert is Death Valley National Monument with its many peculiar and truly beautiful geological features, ranging from salt-encrusted playas and drifting sand dunes to the multicolored canyons and austere mountains which on all sides hem it in. It is a veritable outdoor museum for geologists, and offers unlimited subjects for artists and photographers. The best months for visiting Death Valley are late October, November, and December, when warm, clear, calm weather is the rule. January, February, and March are good months, too, but winds are then more prevalent. If there have been winter rains, snow may tip the higher mountain peaks and wild flowers of many unique kinds may cover the ground in favorable sites. A total of over 600 species of native plants are credited to the Death Valley National Monument. Six species, such as the death valley sage (*Salvia funerea*), napkin-ring spiderwort (*Boerhaavia annulata*), yellow-flowered gilmania (*Gilmania luteola*), holly-leafed spurge (*Tetracoccus ilicifolia*), rock midget (*Mimulus rupicola*), and large-flowered sunray (*Enceliopsis covillei*) are found nowhere else.

The Pacific Coast Borax Company has installed an exhibit of varied objects of unusual historic and natural-history interest at the Furnace Creek Ranch. Here, in one of the company's original buildings and outdoors in the area immediately to the west, are shown many original documents, photographs, books, old vehicles, miners' tools, etc. There is a beautifully lighted hall of minerals; there are also labeled specimens of Death Valley plants and animals. This small but excellent museum enables one to acquire in a few hours more information about Death Valley than could be gained by ordinary methods in many days or months.

On the east, Death Valley is surrounded by the Grapevine and Funeral ranges and on the west by the Panamints, which culminate in Telescope Peak (11,045 feet). Telescope Peak, reached by a road through Wild Rose Canyon to Mahogany Flat and then by trail for 7 miles, offers a view at once magnificent and comprehensive of the greater part of the Mohave Desert. The trip to the lonely peak is one of the most rewarding climbs in the whole region.

The Mohave River arises in the San Bernardino Mountains and flows north and northwestward 147 miles across bleak wastelands, receiving not a single tributary along its length. It is sometimes referred to as an "upside-down river," because for over much of its course the water flows not above but beneath the sandy river bed. Only once in several years is there a continuous flow along the river's length. In places along the river margins are beautiful groves of cottonwood trees (*Popluus fremontii*) and "orchards" of screwbean mesquites (*Prosopis pubescens*, 224).

A railroad and a highway run along the Mohave River for many miles. Since the stream furnished water for man and made available both water and green food for his beasts of burden, the course of the river in early days served as the main trail for hunters, trappers, gold seekers, and immigrants coming into southern California. It later became part of the famed Santa Fe Trail as well as the Old Spanish Trail.

In its lower reaches, just before entering Soda Lake, the Mohave River passes through a gorge, called Cave or Afton Canyon, where is found some of this desert's most spectacular scenery. On one side of the stream are the thick beds of clays which eventually filled the basin of ancient Lake Manix, which once occupied this area. These clays have been formed by erosion into beautiful pinnacles, sculptured crags, and steep embankments. On the opposite side are steep cliffs and talus slopes of highly colored volcanic rocks. Fray Francisco Garcés passed through this gorge and camped in its streamside "caves" (really just big eroded hollows) as he and his followers went westward in 1776 to visit the Mission San Gabriel near Los Angeles. He was the first white man to travel the whole length of the Mohave River. Today a railroad passes through the entire length of Cave Canyon. Auto travelers cannot drive through the canyon, but by using the road south from the point on the highway called Afton one can reach the upper end, from which it is but an easy walk down the canyon.

The highway from Las Vegas to San Bernardino runs for a considerable distance along the Mohave River, and at Baker it crosses Soda Lake, often designated as the Sink of the Mohave because this is as far as ordinary flood waters of the Mohave River ever get. Here, if there is sufficient rain, the water spreads out over a large shimmering clay- and salt-encrusted lake bed but soon evaporates in the hot dry desert air. There is an ancient natural drainage-channel between Soda Lake and Silver Lake to the north, and occasionally there may be enough overflow of river water to fill the Soda Lake playa to the point where it can pour on northward into Silver Lake.

Along the southern border of the Mohave Desert lies Joshua Tree National Monument, an upland area of nearly half a million acres, where typical Mohavean scenery prevails, from sere shrub-dotted basins to low rugged piñon-covered mountains. Here Joshua tree forested valleys are preserved in their primitive state. The Monument features lovely palm oases at Twentynine Palms, 49 Palms, and at Lost Palms. In the Wonderland of Rocks section are huge natural rock-gardens. On clear autumn and winter days from mile-high Salton View (sometimes called Keys View after Bill Keys, who first pioneered a road to this overlook) inspiring vistas can be had of the Salton Sink as far south as Signal Mountain and the rugged Cocopah Mountains in Mexico. From this point of unusual vantage one can also look directly below into the Coachella Valley or view to the south the high ramparts of the Santa Rosa and San Jacinto mountains. To the west lies San Gorgonio Pass, flanked on either side by San Jacinto Peak (10,832 feet) and San Gorgonio Peak (11,485 feet). This pass is often called the Gateway to the Colorado Desert. It is sometimes also called the Pass of the Winds because of the constant winds which blow desertward through it.

Entry to the Joshua Tree National Monument may be made either from the north at Joshua Tree or Twentynine Palms, where the Monument Headquarters are located, or from the south at a point a short distance west of Shaver's Summit.

The animal life of the Mohave Desert is much like that of the contiguous Colorado Desert to the south. The desert tortoise (*Gopherus agassizi*), often erroneously called a "turtle" in spite of the fact that it has clawed toes for walking instead of flippers for swimming, is much more abundant on the Mohave. In the Nevada Desert Game Refuge, north of Las Vegas, there are notable concentrations of the desert bighorn sheep (*Ovis nelsoni,* 134), fortunately now protected so that they are slowly increasing in number to the point where they are no longer in danger of becoming extinct, as once was threatened. These most noble of all desert big game are curious but shy, and are rarely seen at close range. The ewes have more or less straight horns and superficially resemble goats. People who claim to have seen mountain goats on the desert have actually seen female bighorns instead. There are no wild goats in the arid Southwest except perhaps domestic animals that have been released or escaped and gone wild. The bighorn rams have magnificent curved horns that may weigh almost one-tenth of the total weight of the animal. In some cases there is

Philip Lohman

Typical yucca-grassland area of the eastern Mohave Desert with low mountains of volcanic origin beyond. The grass is the galleta grass (*Hilaria rigida*), which here grows between fine specimens of the Mohave yucca (*Yucca schidigera*), deer-horn cactus (*Opuntia echinocarpa*), and barrel cactus (*Echinocactus acanthodes*).

Philip Lohman

Many parts of the high Mohave Desert carry a heavy cover of low shrubs, especially in the well-drained washes of deep granitic sands and gravels. The yucca is *Yucca schidigera*; the low bush in front is the paper-bag bush (*Salazaria neo-mexicana*).

Philip Lohman

Left: The tree yucca (*Yucca brevifolia*) is the most typical tree of the Mohave Desert. This scene was taken in the Joshua Tree National Monument, which was set aside to preserve this splendid "tree lily."

Left, below: Single-leaf piñon (*Pinus monophylla*) a hardy desert pine of the higher desert mountains, often associated with yuccas, junipers, and cacti. The nuts were a valuable source of food to the desert-dwelling Indians.

Below: Nest of the verdin, placed as usual in thorny shrub. Mohave Desert.

The Mohave yucca (*Yucca schidigera*), widespread not only on the Mohave but also in adjacent deserts. Most common in the higher brushlands. Its fibrous leaves were prized by the Indians for making cordage, cloth, and sandals.

Galleta grass (*Hilaria rigida*) and creosote bush (*Larrea divaricata*) on Mohave Desert sand flats. Barren mountains of deep-red volcanic rocks in background.

Ancient lake-bed clays at the lower end of the Mohave River. At the top is a capping layer of black lava marking an era of later volcanic activity.

as much as 30 pounds of horn and head alone! During the rutting season, the rams charge and batter one another, and the crashing of their horns sometimes can be heard at considerable distances. Suprisingly few are ever seriously hurt in these encounters. The sheep of the Desert Game Refuge are now protected both by their rugged, almost inaccessible environment and by the intelligent work of vigilant game wardens who continually patrol the area while making scientific observations on these timid but wary creatures.

The desert "chipmunk" (*Ammospermophilus,* 118), so commonly seen darting across the highway or perched high while feeding on the fruits of a cactus or Joshua tree, is not a chipmunk but a true ground squirrel. In chipmunks the side stripes run to the tip of the nose but in the ground squirrels the stripes end at the shoulder. This nervous but spry little rodent is also called the antelope "chipmunk" in allusion to the white on the undersurface of its tail (like an antelope's) which it continually twitches or holds pressed against its back. Another common rodent of the Mohave is the small bright-eyed, pale-gray, round-tailed ground squirrel (*Citellus tereticaudus,* 117), often seen running across the highway or sitting up straight and motionless like its more northern and larger relative, the picketpin. It has no stripes and its pencil-size tail, kept extended to the rear, is without long fur.

A bird to be eagerly looked for in Joshua tree country is the handsome Scott's oriole (*Icterus parisorum,* 112), finest singer of all the Mohave Desert's birds. Its color is a striking combination of black and deep lemon-yellow, its song a combination of clear rich notes of meadow lark quality. During the spring nesting season the male is heard singing throughout the day. Generally he will be spied sitting at the very tip of a Joshua tree or flying in undulating sweeps from one yucca to another. The nest, semipendant and rather difficult to locate, is woven of yucca fibers and fastened to the yucca's daggerlike leaves.

Both the LeConte thrasher (*Toxostoma lecontei,* 89) and sage thrasher (*Oreoscoptes montanus*) are known to nest in the Mohave's brush-dominated areas. The raven (*Corvus corax*), which is rather scarce on the Colorado Desert, is here plentiful and often seen during sunny days along the highway feeding upon the carcasses of rabbits killed the night before by speeding cars. Few crows ever invade this desert, so if a large black, crow-like bird is seen, it is probably a raven.

In autumn, notable concentrations of migrating turkey vultures or buzzards (*Cathartes aura septentrionalis,* 69), those valuable scavengers of the desert, appear along the Mohave River in large flocks in preparation for their southward migration.

The lizard and snake fauna is quite similar to that of the Colorado Desert. There are several lizards which habitually are associated with the Joshua trees. Several kinds of black rough-scaled lizards (*Sceloporus,* 41) look for insects among the leaves of living trees and the gentle little night lizard (*Xantusia vigilis*) hunts termites in the decaying wood or under leaves of fallen limbs. The most common of the larger Mohavean rattlesnakes is the Mohave rattler (*Crotalus scutulalus*).

The sidewinder or horned rattlesnake (*Crotalus cerastes*) is widespread in the low sandy basins and on the near-by gravelly slopes. It is called the sidewinder because most of the time it glides along at an angle, engaging in a specialized rolling motion that leaves behind a series of disconnected J-shaped tracks. This type of locomotion enables it to move readily up and over steep slopes of loose sand. It is sometimes called the horned rattlesnake in allusion to the prominent pointed scale found over each eye. Oddly enough, the horned viper (*Cerastes*) of North Africa, although not at all closely related, has similar horned scales over its eyes and also moves sideways! This is a curious case of what biologists call "convergent evolution," no doubt induced by a similar sandy habitat. The sidewinder is a true rattlesnake and its venom is as poisonous as that of other rattlesnakes. It rarely exceeds two feet in length. Unlike most other "rattlers," which feed only on warm-blooded rodents, it eats both warm- and cold-blooded creatures, devouring lizards and small snakes as readily as mice.

In the late spring there are often heard in bushy areas the strange crackling sounds of thousands of cicadas or so-called "locusts" which have just emerged from their pupal cases deep in the ground. The noise sounds very much like the crackle of a brush fire and until the insect origin is known may cause considerable apprehension. Other stridulatory insects, such as the desert clicker (*Bootettix argentatus*), which is a small silver-spotted grasshopper found only on creosote bushes, and the large, beautifully marked, tree-yucca-inhabiting shield bearer (*Aglaothorax*), make odd wispy clicking sounds throughout the day, and sometimes into the night, if it be warm. Generally these insects are noisiest in the early morning just after the sun has warmed things up well.

It comes as a great surprise to many to learn that several species of small land snails are to be found living beneath the rock debris and surface rocks of many of the desert mountains. Even the very dry ranges of the Death Valley area have many different kinds represented. The empty and often beautifully brown-banded shells may be found along the margins of the dry hillside or canyon streamways. These small, highly specialized mollusks actively feed only during a few nights or overcast days when tender green plant parts are available. During the long intervening dry periods they hide themselves away in deep crevices or under thick rock ledges or rock slides. They secrete a protective non-porous membrane or epiphragm across the opening of their shell and, sheltered beneath rocks from excessive heat and consequent dessication, await the next wet period when the "seal" can be quickly dissolved and they can emerge and become active for another brief time.

As for man himself, there were two Indian cultures in the Mohave Desert region, the Shoshonean and the Mohavean.

Most of the Indians formerly inhabiting the open Mohave Desert, at least within historic times, belonged to the Shoshonean division of the Uto-Aztecan linguistic stock. In the barren mountains and bleak valleys of the Death Valley region from Owens Lake to and including the Funeral Mountains, dwelt the Koso people, sometimes called the Panamints. Along most of the lands bordering the Mohave River dwelt the Serrano and closely related Vanyume; to the east, north, and southeast, as far as the borders of Mohave Indian territory, were the Chemehuevi. The desert oasis at Twentynine Palms was occupied by Chemehuevi and by Mountain Serrano, who lived mostly in the near-by San Bernardino Mountains. Southern Nevada was occupied by the Paiutes. The localized Mohave Indians were of Yuman linguistic stock and lived closely along the Colorado River, chiefly on the east bank. The name Mohave is from the native word "hamakhava," signifying "three mountains," and referring to the "Needles," prominent landmarks on the Colorado River.

All of the Shoshonean desert territories away from the Colorado River were very thinly populated, and probably not more than 2,500 individuals occupied the entire area. Their food consisted largely of numerous seeds, particularly those of desert sand grass (*Oryzopsis hymenoides*) and chia (*Salvia columbariae*, 328), piñon nuts, mesquite beans, sun-dried prickly pear joints, Joshua tree buds roasted on the open fire, and greens. Animal food, such as rabbits, wood rats, lizards, and some birds were only occasionally taken. These tribes were neither great hunters nor agriculturists, but only simple food-gatherers. Like most of the desert Indians they wore no clothing, except in winter when cleverly made rabbit-skin blankets were used. In the early days rabbit drives were undertaken by entire villages, and thousands of the hapless creatures were slain with rocks and throwing-sticks. The meat was smoked and dried and the hides were cut into strips, then twisted into lines, and later sewn or woven into blankets. To protect their feet from cactus thorns and rough rocks, these Indians wore sandals of yucca fiber or moccasins of buckskin (when they were fortunate enough to kill a deer or desert sheep).

Most of the tribes wove baskets of reeds and grass stems, and some examples of this now lost art are rated as among the finest baskets made in the United States. Little pottery, most of it undecorated, was made, for fragile articles were not desirable for nomadic peoples who must be ever on the move for food. The energetic, inquisitive Mohave people who dwelled along the bottom lands of the Colorado River both on the Nevada-California and the Arizona side belonged to the Yuman linguistic family. They had the most advanced type of culture of all of the Mohave Desert Indians. This was evidenced by their wide travels, their log and thatch houses, their flooded-land agriculture, their various utensils and pottery, and by their numerous games, peculiar clan systems, religious rites, elaborate war pattern, and the way they thought of themselves as a nation, "always thirsty for adventure and glory." Only tribes that practice some form of agriculture which furnishes easy subsistence can become sufficiently sedentary to develop such a high culture.

Each year in May and June the Colorado River, swollen by melting of snow in the high lands of its sources, flooded large stretches of bottom lands, coating them with quantities of soft fertile mud. In this well-wetted land the Mohave planted their crops of beans, melons, wheat, and corn. The summer heat brought their produce to quick maturity. "The relation of the tiller to his strip of fertile soil in the vast burning desert is similar," says Dr. A. L. Kroeber, "to that which obtained in primitive Egypt, and gives Mohavean agriculture a character unique in native America."

Only occasionally are any of these Mohave Desert Indians recognized today, for they wear no distinctive costumes, neither do they live in separate picturesque villages as do other more easterly tribes, such as the Hopi. At Needles, California, the broad-faced, black-haired Mohave and Chemehuevi are sometimes seen walking the streets or selling trinkets at the railway station.

The low-lying, north-and-south–trending Mohavean basins, many of them with strong concentrations of salts in their bottoms, have served as age-long pathways for the southward migrations of plants from the Great Basin Sagebrush Desert, and for the northward spread of many species from the Sonoran Desert. Here, then, we see a strange commingling of two distinctive floras. Only about one-fourth of the plants found on the Mohave may be classed as "endemics," that is, plants that are wholly confined to the area and not found elsewhere. The most specialized and at the same time most restricted in range of these endemic plants are those found in Death Valley, a region which has probably existed as an area of exceedingly low rainfall and high temperature much longer than the desert mountains and arid basins that surround it.

The Mohave Desert wild flower season comes a month or two later than that of the low Colorado Desert. Late March, April, and in the higher elevations even early May may be the time when the flowers, especially the annuals, are at their best. Notable are the stands of squaw cabbage (*Streptanthus inflatus*, 316), the broad fields of royal desert lupine (*Lupinus odoratus*, 354) with deep blue flowers, and the acres of golden gilia (*Gilia aurea*, 248) which convert the usually drab plains and hillsides into places of greatest beauty. Other plants which seem to be especially adapted to the Mohave are the handsome yellow-flowered prince's plume (*Stanleya pinnatifida*, 204), the Mohave aster (*Aster abatus*, 314), the shrubby Mohave sage (*Salvia mohavensis*), the large-flowered yellow sunray (*Enceliopsis covillei*) known only from the Panamint Mountains, the exceedingly prickly-leafed sting bush (*Eucnide urens*), and the gorgeously colored Kennedy mariposa lily (*Calochortus kennedyi*, 284). Except for the Joshua tree and the desert willow (*Chilopsis linearis*, 297) trees are poorly represented. A few smoke trees (*Dalea spinosa*, 348) grow as far north as southern Death Valley. The Mohave yucca (*Yucca schidigera*, 165) is widespread and in the eastern part of this desert the related low-statured big-fruited yucca (*Yucca baccata*) is present.

Travelers crossing the Mohave Desert in August or September may occasionally see flat areas colored bright yellow with the flowers of a low annual called pectis (*Pectis papposa*, 236) When crushed the foliage has a strong but somewhat pleasant, turpentine-like odor. This little plant is among the peculiar desert plants whose seeds germinate only after summer rains, when temperatures are high. It is probably an immigrant from areas with more frequent summer rains. Another pretty desert plant having similar seed germination requirements is the beautiful fringed amaranth (*Amaranthus fimbriatus*), with purple herbage and inflorescence. It is often seen after summer rains in the Joshua Tree National Monument, although by no means confined there, since it is widespread from eastern Utah south and southeastward to Arizona and western Texas; it is found too in Baja California.

The striking Joshua tree is the largest yucca in the United States. It occurs in middle-western Arizona as an associate of numerous small shrubs including the thorny canotia (*Canotia holacantha,* 163) and ribbon-leaved bear grass (*Nolina bigelovii*); in places it there consorts with the giant cactus or sahuaro, which here reaches its northern limit of distribution. In northwestern Arizona, in southern Nevada, and southeastern California the Joshua tree is associated with creosote bush and true sagebrush, and at higher elevations with junipers and single-leaf piñon. This peculiar tree lily, for that is what it really is, grows best on the deep, porous, loamy soils of broad valleys and on the gentle alluvial slopes and pediments surrounding the desert mountains. The plant has no tap roots but instead has numerous corky-barked fibrous rootlets fanning out in all directions to serve both as efficient anchors and for the absorption of water and soil nutriments. The stout trunk, covered with thick, checkered, gray bark, usually sends out branches after reaching a height of five or six feet. These branches and their many divisions bristle with numerous dagger-like saw-edged leaves and eventually at the terminus bear large clusters of creamy-white flowers. The old trees may reach a height of 20 or even 35 or 40 feet. Since it usually is the only sizable tree around, many birds—from hawks to owls, shrikes, and thrashers—build their nests on top of or among the Joshua tree's thorny leaves.

Near Cima and again in close-by Lanfair Valley is found the largest concentration of yuccas in the United States. It is a giant forest of tree yuccas of the subspecies *jaegeriana* (*Yucca brevifolia jaegeriana*), with trees in places so thickly set that it is difficult to see far among them. This yucca differs from the tree yuccas of the western and northern Mohave Desert in having an average shorter main trunk, shorter leaves, a greater tendency to freely branch, and a different chromosome constitution.

In the broad valleys of the northern Mohave several different species of low shrubby plants become dominant; among these are the salt bushes such as the thorny, woody shadscale (*Atriplex confertifolia,* 151), the spiny hopsage (*Grayia spinosa,* 337), and the winter fat (*Eurotia lanata,* 152).

Cacti are poorly represented in numbers except on well-drained salt-free detrital fans or in the lower mountain canyons. Bigelow cholla (*Opuntia bigelovii,* 199), deer-horn cactus (*Opuntia echinocarpa*), spiny-fruited cholla (*Opuntia acanthocarpa*), beaver-tail cactus (*Opuntia basilaris,* 308), Engelmann cereus (*Cereus engelmannii,* 299), barrel cactus (*Echinocactus acanthodes,* 200), and the Mohave niggerhead (*Echinocactus polycephalus,* 201) are found most frequently on the well-drained slopes. On the mountain sides and in rocky canyons grow the pancake cactus (*Opuntia chlorotica,* 198), the Mohave opuntia (*Opuntia mohavensis*), hedgehog cactus (*Opuntia erinacea*), and a number of low-growing nipple cacti (*Mammilaria,* 351).

13

THE *Great Basin* DESERT

The Sonoran Desert comes to a rather abrupt northern end where it reaches southern Nevada, midwestern Arizona, and mideastern California, north of Death Valley. To the north lies the largest of all American desert areas, the GREAT BASIN DESERT, an arid upland country made up of intermountain plateaus, valleys, and broad basins shut off from the Pacific's moisture-laden sea-winds by the High Sierra and other lofty mountains bordering it on the west. In it are included the "enclosed-basin deserts" of northeastern California, eastern Oregon, and most of Nevada and Utah, the Red Desert of Wyoming, and some arid parts in the drainage basin of the Snake River of southern Idaho. (See map, p. 124.)

In the Great Basin Desert the green of the creosote bush, so characteristic of the southern deserts, is displaced by the soft gray of a number of dominant low-statured shrubs. Among the commonest of these is the well-known toothleafed sagebrush (*Artemisia tridentata,* 155), often popularized in Western fiction and now the state flower of Nevada. Because of the abundance of this shrub in many parts of this region this is called the Sagebrush Desert, and sometimes, in reference to its scientific name, it is termed the Artemisian Desert.

Vying for attention with the sagebrush is the oval-leaved shadscale or sheep fat (*Atriplex confertifolia,* 151), a low, somewhat spiny, woody-stemmed salt bush found widespread on alkaline plains and mesas of arid North America from eastern Oregon, Montana, and South Dakota southward to the Chihuahuan Desert of Mexico. In the Great Basin it is often especially prominent where it forms almost pure stands and, in monotonous sequence, crowds out nearly all other plants over vast areas of the alkaline basins. Recent surveys show that in Utah it probably covers as much or more land than the sagebrush. The plants are usually well rounded, compact, and seldom over one or two feet high. The tender shoots, crowded with almost-white, mealy-surfaced leaves, are an important forage for livestock. Shadscale was one of the shrubs described by Dr. John Torrey and Captain John C. Frémont after Frémont's memorable exploratory journey through northern Utah in 1843.

At lower elevations throughout this area also grows such drought-resistant plants as hop sage (*Grayia spinosa,* 337), and mule fat, sometimes called winter fat (*Eurotia lanata,* 152).

On many extensive, low flat areas where moist soils are prevalent we find only such hardy, small-leaved, alkali-resistant shrubs as the true greasewood (*Sarcobatus vermiculatus,* 139), especially prominent on the Great Salt Lake and Sevier deserts of Utah, gray Molly (*Kochia vestita,* 138), mat salt bush (*Atriplex corrugata*), pickleweed (*Allenrolfea occidentalis*), samphire (*Salicornia rubra* and *S. utahensis*), and the salt grass (*Distichlis spicata*). The true greasewood (*Sarcobatus vermiculatus*), just mentioned, should not be confused with the creosote bush (*Larrea,* 238) of the Sonoran Desert, which in Arizona and California is often erroneously called "greasewood"; the two plants are distinctly different and ordinarily never occur together. The true greasewood is a gray-green, succulent-leafed plant of saline moist soils, whereas the creosote bush is a brownish-green, thin-leafed plant of the lower, largely alkaline-free dry plains and mesas.

Two kinds of desert teas (*Ephedra nevadensis* and *E. viridis*) are scarce to abundant in rocky hilly areas of the Great Basin Desert. Over some of the large flats the yellow-flowered, late-summer-blooming rabbit brushes (*Chrysothamnus puberulus* and several other species) dominate the vegetation to the exclusion of all other plants. Rabbit brush is a shrub which often, like the matchweed (*Gutierrezia*), takes over in places which have been burned or overgrazed.

In contrast to the more southern deserts, cacti are here a relatively unimportant feature of the flora. The most common species occur in scattered clumps and include a few low-growing kinds of opuntia, which are able to withstand long periods of growth inactivity during the cold winters and long dry summers. In arid portions of eastern Oregon and Washington, and along the Snake River in Idaho, grows the widespread Great Basin cactus (*Opuntia polyacantha*), a species known to range as far as 500 miles north of the Canadian border. Another cactus, *Opuntia fragilis,* is found even in the Yukon area.

The very prevalent sagebrush (*Artemisia tridentata,* 155) is easily identified by its spatulate gray-green leaves bearing three short wedge-shaped teeth at their tips. The plant is not a true sage but a variety of wormwood, and only received its common name of sage because of the pungent, sweetish, sage-like odor of its crushed foliage. The sage hen (*Centrocercus urophasianus*), which in winter feeds almost entirely upon sagebrush leaves, is not too popular as food because of the strong "sagey" taste of its flesh. When growing on well-drained alkaline-free gravelly loams, the sagebrush grows to a height of two or three feet, under such conditions often forming pure stands over large areas. In the mineral-rich soils formed through the disintegration of lavas, such as are found in southern Idaho and eastern Oregon, the sagebrush develops not only a very extensive root system but extraordinarily large thick stems and may reach the height of a horse's back. The dry wood of this plant burns with a very hot flame and gives off a pleasant-odored smoke. The bitter foliage is sometimes heavily grazed by livestock. When cultivated as an ornamental in gardens, it may make a most attractive showing.

The Great Basin Desert is for the most part a broad and arid land of rather bleak and monotonous scenery, devoid of trees except along the few streamways where willows and cottonwoods occur. Most of it is fairly high country, with elevations ranging between 2,000 and 5,000 feet or more above sea level. Between the more or less parallel ranges of high, steep, thrust-fault or fault-block mountains, which traverse the entire area, lie scattered low, barren or piñon-and-juniper-covered ranges and numerous broad sandy or gravelly valleys and depressions. In the lowest depressions are salt- and alkali-encrusted playas or dry lakes, often of large size; none of them have drainage outlets to the sea. Some of the barren salt flats are so hard-surfaced, so nearly level for long distances, and extensive, that they have become well known as automobile "race tracks" used especially to establish speed records. Al Jenkins and the British Sir Malcolm Campbell have made famous the salt flats near Wendover, Utah, on the western edge of the large, 100-mile-long "Great Salt Lake Desert." It was the vastness, heat, and sterileness of the surface of this same desert, called the Seventy Mile Desert by Captain Howard Stansbury, that was the cause of one of the worst tragedies of the Donner Party as they passed over it on their way to California in 1846. "After two days and two nights of continuous travel over a waste of alkali and sand," wrote Elizabeth Donner Houghton, one of the last survivors of the party, "we were still surrounded as far as eye could see by a region of fearful desolation. The supply of feed for our cattle was gone, the water casks were empty, and a pitiless sun was turning its burning rays upon the glaring earth over which we still had to go." During this September crossing of this desert many of the party's cattle died from thirst and exhaustion, and the lot of the immigrants became more alarming every day. To add to their misery the Indians made a raid upon the cattle and killed or drove off twenty of their animals.

In a number of parts of the Great Basin Desert, such as the Snake River plains of Idaho and in portions of southern Oregon, there are picturesque cones of black or red volcanic cinders, extensive black and barren lava flows, and upraised blocks of banded basalt.

Like most regions of major aridity, the Great Basin has its areas of sand dunes. Probably the largest of these is found north of Winnemucca, Nevada. These dunes average about 75 feet high. They vary in width from 8 to 10 miles and are about 40 miles long. They are gradually migrating eastward. Another considerable area of dunes lies south of Carson Sink, where the sand lies 200 to 300 feet thick.

In many ways, much of the Great Basin is the least spectacular of all our Southwestern desert-lands. But few regions exist which do not have some attractions. In this particular desert, the strong appeal to every lover of wild places lies in the very vastness and seemingly endless expanse of its broad, pebble-strewn and brush-covered basins, and in the surrounding bizarre, sere, and often barren but colorfully banded mountain ranges, which rise with remarkable steepness from the desert floor.

Mountains of sufficiently high elevation to support forests of coniferous trees are widely scattered, and during the winter and early spring they are generally covered with snow. When climbed, these high ranges offer magnificent panoramic views of the surrounding sun-drenched, brush-carpeted basins and interspersed, often extensive, ghostly-white alkaline flats.

Such isolated mountains are true biologic islands in a sea of aridity, harboring on their summits a peculiar boreal biota. The surrounding desert is just as much a barrier to the spread of the animals and plants living on these mountain tops and slopes as are the ocean waters to life on a coral atoll.

Of rivers there are but few. In Nevada the largest is the Humboldt, which arises in the north and winds tortuously southwestward through deep gorges it has cut into the north-south ranges crossing its path. Eventually it empties its waters into an undrained lake basin called the Humboldt Sink. Other smaller rivers are the Quinn and the Kings, which vanish in the broad alkali flats of the Black Rock Desert; the Muddy and the Virgin, which drain into the Colorado; the Truckee, which feeds Pyramid Lake north of Reno; and the Walker, which empties into Walker Lake. The Carson River, which arises in the western mountains, now flows for the most part into the Lahontan Reservoir; formerly it discharged all of its water into the large Carson Sink, which covers approximately 100 square miles and whose broad marshes and shallow lagoons today attract notable multitudes of water birds, such as the glossy-faced ibis, the red-winged blackbird, the coot or mudhen, and many kinds of ducks and geese.

In Utah the longest stream of the parched desert domain is the Sevier River, which rises in the south-central high plateau and runs circuitously north, then south, to drain into Sevier Lake, a large body of saline water of the central-western unreclaimable shadscale desert where cattle raising has become the main industry. Often Sevier Lake is dry because the water that formerly ran into it is diverted for irrigation before it reaches the lake bed. Flood waters may occasionally flow into it.

The vast upland area of the Great Basin was in late glacial and postglacial times a land of many inland streams and of chains of large pluvial lakes, filled by water from melting snow and glacial ice, and an increased rainfall. Geologists call the two largest of these prehistoric lake-chains and feeder streams the Lahontan Drainage System of Nevada,* and the Bonneville System of Utah.† Great Salt Lake is a small remnant of the latter system. Other Nevada lakes and streams—and there were many indeed—formed, among others, the large Railroad Lake System of the western and central parts of the state. The Pahranagat Lakes probably drained through the large tortuous Pluvial White River of eastern Nevada into the lower Colorado River.

Highly saline Great Salt Lake, estimated to contain some 400,-

* Named after Baron Lahontan, a French soldier who traveled in America in the late seventeenth century and wrote an account of his discoveries. In his book he mentions a "great salt lake," knowledge of which he gained from the accounts of Indians. See *Exploration and Survey of the Great Salt Lake of Utah*, by Howard Stansbury (1852).

† Named in honor of Colonel Benjamin E. Bonneville, once commanding officer of Fort Kearney and of the Gila Expedition in 1857. His explorations in the Rocky Mountains and California (1831–36) were published in a journal which later was amplified by Washington Irving and published under the title *Adventures of Captain Bonneville*.

000,000 tons of common salt and 30,000,000 tons of sodium sulphate, Utah Lake (which is fresh water only because it has an outlet into Great Salt Lake through the Jordan River), and saline Sevier Lake (Lake Salada of the old maps) to the southward, into which Sevier River drains, are but small shrunken relicts of the huge ancient Lake Bonneville whose waters rose to levels as high as 1,000 feet above the present Salt Lake floor (4,200 feet). Clearly seen wave-cut cliffs and wave-built terraces, dry deltas, "fossil sand bars," and beaches occur at several levels on the flanks of mountain ranges of northwestern Utah, marking the old strandlines of this enormous lake that at one time covered an area of about 20,000 square miles and had an extreme depth of about 1,050 feet. The first high-water stage of Lake Bonneville was followed by a long period of drought when the lake almost, or actually, dried up. This event was followed by a new high-level filling, bringing the water 90 feet above the previous strandline and 1,000 feet above the present Great Salt Lake. The record of this is preserved by the Bonneville Strandline. The waters emptied through an outlet at the lake's northern end, called Red Rock Pass, into the Snake River. Wearing down of this outlet to a hard ledge of bedrock soon reduced the lake level to 4,825 feet (625 feet above the present Great Salt Lake). This level was maintained for a long time and the Provo Strandline was established. Later evaporation caused further shrinkage of the lake and another lower strandline, Stansbury Strandline, was developed at 4,530 feet (330 feet above the present lake).

Honey Lake in California, Pyramid Lake, Walker Lake, and the Carson and Humboldt sinks in Nevada are residual parts of the gigantic prehistoric Lake Lahontan, which once covered 8,400 square miles. Many Nevadan valleys once connected by surface-stream waters with Lake Lahontan are now completely cut off by intervening divides. Evidence of their former connection can be found in the small relict populations of Cyprinodont minnows found now only in some of the isolated springs of the area but which were once widespread and probably occurred in great numbers in lakes and streams over the entire region.

Pyramid Lake, "set like a gem in the mountains," north of Reno, Nevada, is a body of only slightly saline waters, about 30 miles long and 7 to 10 miles wide. The three distinct terraces seen at 110, 320, and 530 feet above this lake mark former levels of ancient Lake Lahontan of which Pyramid Lake is but one of the shrunken remnants. Jutting up from the water of this beautiful lake are a number of rugged and picturesque islands of eroded tufa. One of these, rising nearly 500 feet above the water, is shaped much like the giant pyramid of Cheops, hence the name Pyramid Lake, given to it by Captain John C. Frémont, who visited and described this scenic body of water in the winter of 1844.

Anaho Island near the southeastern end of the lake is now a 248-acre bird refuge, set aside to protect the great numbers of pelicans which live there while feeding at the mouth of the Truckee River during the run of the "land-locked" salmon (the "cui-ui" of the local Paiute Indians). At the northern end of Pyramid Lake, rising abruptly from the blue-green water, is a grotesque cluster of sharp, stone "teeth," really tufa pinnacles, known as The Needles.

In the rocks of the north-south-trending mountains which separate the debris-filled valleys and basins of the Great Basin Desert, lies great mineral wealth. Here are to be found numerous rich mines of gold, copper, zinc, silver, and lead. Some of the extensive salt-encrusted dry lakes or playas yield vast quantities of important commercial salts. It is all a land rich in the lore of mining and miners, of ore strikes and boom towns.

In south-central and eastern Oregon is an elevated near-level land tract about 150 miles long and from 30 to 50 miles wide, comprising what is called on many old maps the Great Sandy Desert, or on recent maps the Harney High Desert. Here and there above its surface rise picturesque buttes and mesas, between which lie wide areas of dust and sand which have been formed by the breakdown of pumice. Beneath much of the loose powdery surface-soil are fissured sheets of lava. There are practically no streams, and what little runoff there is during the wet season finds its way to the bottoms of enclosed basins, there to form ephemeral lakes; these later in the year become white-encrusted playas. To the east of the Harney High Desert on the plateau of southeastern Oregon lies the Malheur or Harney Basin where there are two large but shallow lakes, one called Lake Malheur, the other Harney Lake. These were formed when an obstructing lava block was made in Malheur Gap across the former stream bed of the Silvies River, a stream which once discharged through Malheur River into Snake River, and finally into the Columbia River and the Pacific Ocean. Although usually included in the Great Basin, Lake Malheur and its sister lake, Lake Harney, are really disrupted parts of the Columbia River system. Farther eastward are the Snake River plains of Oregon (not to be confused with the Snake River plains of Idaho) embracing an arid plateau area of 1,200 square miles, which consists of fantastic formations of deeply sculptured basaltic hills, small craters, and intervening valleys. Most of the streams of this area flow through deeply carved, scenic canyons and find outlet into the Owyhee River and thence into the Snake River and the Pacific Ocean. The so-called Alvord Desert, lying within the Basin-Range Province at the eastern base of the Steen Mountains, consists for the most part of a dry-clay-encrusted playa without drainage outlet. After spring rains and snow thaws, its fifty or sixty square miles of surface may be covered with water from a few inches to two feet deep.

Of great interest to students of fossils is Oregon's Fossil Lake in the bed of ancient Fort Rock Lake, a former large pluvial lake of the desert country in Lake County, south of the Great Sandy Desert. Here is found a rich fossil bird fauna, including bones of a flamingo, and mammalian remains such as those of the giant beaver (*Castoroides*), as well as the common present-day beaver (*Castor*). The finding of jawbones of large salmon leads to the conclusion that Fort Rock Lake once had a connection with the Pacific Ocean. The beautiful, well-marked parallel terraces and wave-cut notches in the ancient shoreline rocks of old Fort Rock Lake are striking features of the desert plains north of Silver Lake.

The barren Black Rock Desert of northwestern Nevada, which has an average elevation of about 4,000 feet, contains a huge alkali flat which in spring may turn into a million-acre shallow lake and later into a "morass, slick as grease and completely impassable." Persons visiting this desert wasteland in summer find in its middle only a shimmering, "almost dead-level," sunbaked pavement of smooth, hard, white salt-encrusted clay. Black pyramids of volcanic rock rising above this dry surface "like the dorsal fins of giant sharks" give the name Black Rock to this desert.

Once widely plentiful all over the West, the noble pronghorn antelope (*Antilocapra americana*) still survives in limited numbers in the Charles Shelton Antelope Refuge of the bleak Black Rock Desert of the hinterland of northwest Nevada, and in the Mountain Antelope Refuge of southeastern Oregon and in certain areas of the lowlands of Montana. Because of a scarcity of food during the midsummer months, some of the smaller mammals such as the Townsend ground squirrel (*Citellus townsendi*), which inhabits the lower elevations of the northern Great Basin Desert, become dormant for that period and engage in a kind of torpor quite similar to winter hibernation, but which, because it occurs in summer, is termed estivation (from Latin *aestivare,* "to spend the summer").

The small sagebrush chipmunk (*Eutamias minimus*) is most often found around dense clumps of sagebrush and rabbit brush, and may often be seen climbing in the tops of the bushes gathering or eating the small seeds. Gophers on the desert seem at first an incongruity, but each of the major desert areas has its distinctive pallid-colored species. The Great Basin contains, among others, the typical light-colored gopher (*Thomonys perpallidus*), most abundant in western and central Nevada.

Two kangaroo rats (*Dipodomys ordii* and the rarer *Dipodomys microps*), and the tiny black or brownish, large-headed kangaroo mouse (*Microdipodops megalocephalus*), with tail peculiarly swollen in the middle, occur on the Sagebrush Desert, the latter being restricted to sandy areas. Among the several kinds of mice should be mentioned the white-bellied grasshopper mouse (*Onychomys leucogaster*), the big-eared harvest mouse (*Reithrodontomys megalotis*), the very common and widespread white-footed mouse (*Peromyscus maniculatus*), and the buffy-gray, rock-inhabiting canyon mouse (*Peromyscus crinitus*). The most common pack rat is the widespread desert pack rat (*Neotoma desertorum*). The jack rabbit (*Lepus californicus,* 114) and the slate-gray pigmy brush rabbit (*Sylvilagus idahoensis*), particularly fond of sagebrush thickets, are widespread. Ground squirrels include rock or brush-inhabiting species such as the Oregon ground squirrel (*Citellus oregonus*), the soft-haired ground squirrel (*Citellus mollis*), and the antelope ground squirrel (*Ammospermophilus,* 118). Carnivores include the often beneficial coyote (*Canis latrans*), the badger (*Taxidea taxus,* 129), and the bob cat (*Lynx rufus,* 128).

The number of birds inhabiting this northernmost desert country is considerable. The turkey vulture or buzzard (*Cathartes aura,* 69), and the raven are again the prominent birds of carrion. Among the birds of prey are the sharp-shined hawk (*Accipiter velox*), Cooper hawk (*Accipiter cooperi*), red-tailed hawk (*Buteo borealis,* 68), Swainson hawk (*Buteo swainsoni*), marsh hawk (*Circus hudsonius*), horned owl (*Bubo virginianus pacificus,* 56), long-eared owl (*Asio wilsonianus,* 61), and screech owl (Otus *flammeolus*). The killdeer (*Oxyechus vociferus*) is fairly common, as is also the mourning dove (*Zenaidura macroura,* 93), the Nuttall poorwill (*Phalaenoptilus nuttalli,* 83), and the Pacific nighthawk (*Chordeiles minor*). The white-throated swift (*Aeronautes saxatalis,* 80) wings its way and forages over the desert but nests in the higher mountains near by. The commonest hummingbird of the Great Basin is the broad-tailed hummingbird (*Selas-*

Walter P. Cottam

Wah Wah Valley in the central area of the Bonneville basin of Utah is typical of the numerous valleys of the Great Basin. In the foreground is the desert shrub (*Tetradymia spinosa*). On the upper slopes of the valley immediately below this ridge is a narrow belt of sagebrush (*Artemisia tridentata*), generally more extensive than shown here. Below is a vast expanse of shadscale (*Atriplex confertifolia*). This gives way in the low valley areas to greasewood (*Sarcobatus vermiculatus*) and frequently to barren playa lakes. The Wah Wah Mountains in the distance, like most ranges of the Great Basin, are seen to be buried to their "necks" in their own debris.

Steens Mountain and Sheephead Mountain form a distant skyline far across t[illegible]
tensive acres of the Alvord grazing lands in Harney County, southeastern Oregon. [illegible]
dry, alkaline-incrusted lake in the bottom of the basin.

Oregon State Highway Commission

Walter P. Cottam

An embayment of ancient Lake Bonneville near Great Salt Lake, showing the old shoreline on the mountain in the middle distance, with a playa lake at its base. In the foreground, on the oolithic sandy beach, are narrow belts of rabbit brush, greasewood, and samphire-pickleweed occupying the flat terrain adjacent to the playa bottom. Each belt marks a progressive rise in water table and saline content.

Oregon State Highway Commission

Steens Mountain in southeastern Oregon rises 5,000 feet above the expanses of arid Alvord Valley, which itself is 4,000 feet above sea level. The mountain has a length of more than 40 miles and is from 15 to 20 miles across. Note enormous alkaline flats.

phorus platycercus), with rattling, bell-like notes made in flight by air passing over the attenuated primary wing feathers. It feeds upon the nectar of the flowers of the paintbrush, cactus, and other plants, particularly those with bright-colored tubular corollas.

The commonest woodpecker is the red-shafted flicker (*Colaptes cafer collaris*); the most frequently seen flycatchers are the western or Arkansas kingbird (*Tyrannus verticalis*), and the Say phoebe (*Sayornis saya*, 63). The brownish-gray bird with a black collar and yellow throat, so often flushed from places along the sides of the road where it occupies short, scrubby vegetation such as shadscale, is the horned lark (*Otocoris alpestris*, 86). Overhead may be seen the graceful violet-green swallow (*Tachycineta thalassina lepida*).

In those portions of the Sagebrush Desert where there are piñons and junipers, the piñon jay (*Cyanocephalus cyanocephalus*, 74) can usually be seen in spring, summer, and autumn in small but noisy flocks. Here also come groups of the bushtit (*Psaltriparus minimus*) and of the mountain chickadee (*Penthestes gambeli*). Wherever there are fairly large rocks, the rock wren (*Salpinctes obsoletus*, 102) may be seen bobbing up and down or running about in search of crevices, where it finds spiders to feed upon. The mockingbird (*Mimus polyglottus*, 100) is not common in the Great Basin but may occasionally be seen.

The western gnatcatcher (*Polioptila caerula*) is sometimes seen but again is not as common as it is farther south. The loggerhead shrike or butcherbird (*Lanius ludovicianus nevadensis*, 85) lives in parts of the Great Basin where there are bushes high enough to serve as lookout perches, and sufficient open ground to be surveyed for its favorite food: lizards, small rodents, small birds, and large insects. The black-throated gray warbler (*Dendroica nigrescens*) is the commonest warbler and is especially numerous in the piñon-juniper belt. In the lower valleys of the Great Basin can often be heard and seen the well-known western meadow lark (*Sturnella neglecta*, 111), and around human habitations may be observed small flocks of the ubiquitous English sparrow (*Passer domesticus*). The only common oriole is the bullock (*Icterus bullocki*); the house finch or linnet (*Carpodacus mexicanus frontalis*), so abundant throughout the Mohave and Sonoran deserts, is comparatively rare.

The most frequently seen sparrows include the lark sparrow (*Chondestes grammacus*), black-throated sparrow (*Amphispiza bilineata deserticola*, 104), sage sparrow (*Amphispiza nevaden-*

sis), Brewer sparrow (*Spizella breweri*), and song sparrow (*Passerella melodia fallax*).

There are three birds particularly associated with the sagebrush: the sage grouse (*Centrocercus urophasianus*), the sage thrasher (*Oreoscoptes montanus*), and the sage sparrow (*Amphispiza nevadensis*, 103), with its exquisitely sweet song, particularly noticeable during the breeding season. Both the sage thrasher and the sage sparrow usually build their nests in sage bushes, utilizing fine shreds of bark as nesting material. Writing of the sage thrasher, Florence M. Bailey, a well-known student of western bird life, said: ". . . his commonest perch is the top of a tall sage bush, and as his song is poured out even long after dark and sometimes by moonlight, with scarcely less richness than the true thrasher's, you are glad he lives in the deserts." Like other desert thrashers, the sage thrasher has a way of running rapidly from bush to bush, thus avoiding easy capture by avian enemies.

Here the commonest amphibian is the spade-foot toad (*Scaphiopus hammondi*), a strictly desert inhabitant which shows considerable range in color, from pale gray to dark green. It receives its common name because it employs its hind feet, which are equipped with hard, sharp "spades," for digging into the mud, where it remains through the dry portions of the year, emerging usually only after a heavy rain. Until summer water holes begin to dry up, the spade-foot toad forages abroad both in daylight and night hours; as it becomes hotter and drier, activity is restricted to the night hours. During the day it retreats into burrows which it digs in the mud about desert water holes, small streams, and temporary pools.

Great Basin reptiles include the large carnivorous collared lizard (*Crotaphytus wislizeni*), which is unique among lizards because the female is the more brilliantly colored. During the breeding season her sides take on beautiful brilliant hues of salmon pink, accentuating the dark markings which suggest the common name, leopard lizard. The western collared lizard (*Crotaphytus collaris baileyi*, 42) is not as common here as further south and is apparently restricted to areas with fairly large rocks. The little brown-shouldered lizard (*Uta stansburiana*, 37) lives among the rocks and sagebrush, where it seeks out its insect prey. The mountain swift or sagebrush lizard (*Sceloporus graciosus*) is fairly common and confined mostly to sagebrush areas. There are two species of horned lizards of the desert area, *Phrynosoma platyrhinos* (44), and *P. douglasi ornatis*, which inhabits the salt flats around Great

Salt Lake. The whip-tailed lizard (*Cnemidophorus tigris*), with long body and tail and slinking gait, is confined to lowland and desert valleys.

Among the snakes of the Great Basin should be mentioned the very agile striped racer (*Coluber taeniatus*), the gohper snake (*Pituophis catenifer deserticola*), the desert night snake (*Hypsiglena torquata deserticola*), and the Great Basin rattlesnake (*Crotalus viridis lutosus*). Some of these snakes are known to congregate in large numbers in dens for hibernation. Dr. Angus Woodbury of the University of Utah studied for several years one such den in Tooele County, Utah, where about 1,000 snakes get together to winter.

Much of the desert area of the Great Basin was occupied by the Paiute Indians, whose economic life was based upon hunting and seed gathering. Their contact with Europeans probably began with the entrance of trappers to their territory, about 1825. A great crisis to their economy came with the westward tide of white immigration when gold was discovered. Piñon forests upon which they depended for nuts were felled for fuel and timbers for mining operations. Introduction of grazing animals meant destruction of many of their food and medicinal plants.

Fossil fields of great interest are found in the Great Basin Desert. One of the largest of these is the Miocene Esmeralda Field, located in Nye and Esmeralda counties of Nevada. Here mammalian remains are found in deposits along the shores of former fresh-water lakes. Fragmentary remains of now-extinct forms of elephants, bison, camels, and horses have been found in the sediments and gravels along the shores of postglacial Lake Lahontan.

The rocks of the Charleston or Spring Mountain area of southern Nevada contain an abundance of fossils, most of them representing invertebrate animals that lived in seas. The Carboniferous limestones are rich in fossils, but they have no monopoly. All of the Paleozoic systems are well represented from Cambrian to Permian; in addition, the foothills have thick deposits formed in the first two periods of the Mesozoic era. The most important forms in the Cambrian rocks are trilobites (primitive crustaceans). Rocks of the later Paleozoic systems have bivalve shells (clams and brachiopods), gastropods (snails), crinoids ("sea lilies"), and many others. The Mesozoic rocks have fossil reptiles and large petrified trees.

The Red Desert of Wyoming, while properly not a part of the

Great Basin, is, because of its proximity, considered here with it. The name was originally given to a small tract, 15 or 20 miles in extent, known for its peculiar red, shale-derived clays, but the title is now used to designate all of the 11,000 square miles of salt-impregnated soils in southwestern Wyoming where "salt sages" of the genus Atriplex are the dominant plants. It is mostly a lonely, rather monotonous land of an average elevation of 7,000 feet, enclosed by mountains and intersected at intervals by low ranges of hills. Occasional buttes occur; some of them, like Negro Butte, are conspicuous landmarks. Since most of the Red Desert straddles the Continental Divide, some of its drainage is toward the Pacific through the Green River and some toward the Atlantic through the Platte.

To the north of the railroad which transects the area, the country is almost level for 30 or 40 miles. Beyond that it is rimmed in on the north by high and often steep bluffs. In the northwestern part the region is very broken, and here spectacular castle-like buttes rise to a height of 2,000 to 3,000 feet.

The rainfall averages less than ten inches. The short summers are dry, but cloudbursts, sometimes of major proportions, may occur; tortuous Bitter Creek with its minor tributaries serves as the principal means of carrying off the murky, clay-darkened waters. Such waters as normally flow are so nearly saturated solutions of soil-leached salts and minerals that they are never fit for drinking.

The most characteristic vegetation consists of three-tooth sagebrush (*Artemisia tridentata*, 155), and bud brush (*Artemisia spinescens*), mixed with the spiny shadscale (*Atriplex confertifolia*), several kinds of salt sage (*Atriplex*), winter fat (*Eurotia lanata*), grayia (*Grayia spinosa*), greasewood (*Sarcobatus vermiculatus*), and wheat grass (*Agropyron*).

In the spring season, which is the time when heaviest precipitation occurs, a considerable number of colorful annual wild flowers grow in the spaces between the hardy shrubs. But by the end of June they have gone to seed and dried up. The area is used for the most part as a grazing ground for sheep.

14

THE *Painted* DESERT

Lying in elevation just above the creosote bush and the sahuaro deserts is a belt of semidesert land where sturdy piñons (*Pinus edulis*) and low-statured junipers (*Juniperus*) of several species grow, together with perennial sod-forming grasses and shrubs such as the wormwoods (*Artemisia*). Such areas dominated by piñon, juniper, and artemisia are particularly common in northern Arizona, southeastern Utah, and northwestern New Mexico. Much of this steppe-desert is a scenic plateau land of rocky terrain, surrounding low mountains and extensive flat-topped mesas which jut up like colorful islands and headlands from the level desert floor. It includes most of the drainage area of the Little Colorado River, as well as portions of the upper Colorado River and of the San Juan River in southeastern Utah. (See map, p. 124.)

Because of the altitude of this area (3,500 to 5,500 feet), the winters are generally cold, but the summer days may often be quite hot. Snow from time to time covers the ground in winter. Local rains, often of considerable intensity, fall in summer. Banks of beautiful cumulus clouds usually fill the deep-blue summer skies by day. The summer nights are made glorious by the array of brilliant stars or by the display of sheet and forked lightning from distant storm-clouds.

Since much of this desert is typical Navaho Indian country it is sometimes called the Navahoan Desert. The term "Navahoan" is, however, thought to be ill chosen, for the Navaho* are not the oldest and longest established representative Indian group in the area. Not until the late 1700's did they begin to filter in around the

* There is some difference of opinion on the formation of the plural of Indian group names. Many add "s" but this does not solve the plural for a couple of these names. For example, Blackfoot. Many make it Blackfeet for the plural. Perhaps it is best to use the same word for singular and plural. Hopi and Hopi, Navaho and Navaho. Such usage saves a lot of trouble, and can be used consistently.

well-established Hopi and not until the middle 1800's did they settle extensively. The name PAINTED DESERT seems to me to be better because it is more descriptive and comprehensive, and is not in any sense misleading. Some of this desert's best scenery is in the spectacular, Indian-inhabited badlands and mesas which stretch from northern New Mexico across Arizona to the eastern edge of the Grand Canyon National Park.

Highway 66 cuts across some of the most picturesque eastern parts of the Painted Desert at the north end of Petrified Forest National Monument near Holbrook, Arizona; the Monument's rimroad along the old lava flow should not be missed. Highway 89 north from Flagstaff, Arizona, skirts the western edge of this land of grotesquely eroded landscapes. Even more spectacular portions of the Painted Desert can be seen by taking the road northeastward from Cameron, past the famous Elephant's Feet and into Monument Valley lying directly on the Utah-Arizona line. Monument Valley is probably one of the most photographed spots in America today and certainly one of its most distinctly beautiful.

Here the Navaho Indians and their primitive way of living can best be seen. Another great scenic attraction in the Navaho country is Canyon de Chelly National Monument, best reached by going northwest of Gallup, New Mexico. Here in a sheer-walled, red-sandstone canyons still live the seminomadic Navaho. On the cliff walls around them are some of the best-preserved cliff-dwelling ruins found in the Southwest, ruins that date back some 800 or 900 years. Rich rewards indeed are here for those who penetrate this enchanting primeval wilderness. The Navaho Indians, now well-established in northern Arizona and northwestern New Mexico, are renowned for their silversmith work and for their beautiful blanket weaving. They are at present engaged mostly as herders of sheep and goats and as raisers of cattle. In the summer on their small, sometimes temporary, farms they grow maize, beans, pumpkins, squash, and muskmelons. Their corn, often grown without irrigation, is sometimes planted in irregularly placed hills, but also in rows or in circular patterns. For winter occupancy the Navaho build crude but often picturesque huts called "hogans," using mud-covered piñon and juniper logs; often they cover the hogan roof with mud. For summer, only an open framework of green piñon or juniper branches suffices.

Some of the ancestral groups represented among the Navaho came down onto the Colorado Plateau from the north. The Navaho as we know them today really date from the 1600's or perhaps

1500's, when they became a distinct group of the Apache Indians of an area called by the Spanish "Nabahó," an area of northwest New Mexico west of the Río Chama. The group accepted people from various Pueblo tribes, from the Ute and others, and the modern Navaho is the result. The Navaho were peaceful during the early Spanish period but as the Spanish influence expanded they turned to raiding and kept up an almost constant predatory war with the Pueblo peoples as well as the white settlers. A marked revolution in their way of living came about with the introduction of sheep raising during the Spanish period, and later with the occupation of their lands by the United States Government.

In the center of the present Navaho domain is the reservation of the Hopi Indians, one of the several small groups of Pueblo peoples who inhabit this desert. The Hopi built adobe and stone "pueblos" atop the lofty plateaus or mesas of northeastern Arizona, where they could best protect themselves against their often marauding neighbors, such as the Ute and later the Navaho. They are said to be the only Shoshonean people who ever adopted a Pueblo culture. Like other desert tribes, they left few plants unutilized; indeed they found practical or ceremonial uses for almost every species. They have long been expert "dry farmers." Much popular interest has been shown in the spectacular Hopi snake dance held every two years. The colorful public performance is only a small part of a much longer ceremony. The name, Hopi, is one coming from their own name, Hópitu, meaning "peaceful ones."

Running like fingers through much of this high semidesert between the Little Colorado and the Colorado rivers of northeastern Arizona and southeastern Utah are wide areas where the plant cover is very sparse, due either to low rainfall or the rockiness of the soil. In stretches dominated by sandy soils there are often myriads of small wind-heaped hummocks separated by barren spaces. Atop these hummocks may be seen protruding clumps of partially buried desert tea (*Ephedra viridis* and *Ephedra cutleri*) and scattered tufts of the beautiful big-seeded sandbinding grass sometimes called Indian rice (*Oryzopsis hymenoides*). Shrubby yuccas (*Yucca baileyi*, 157, and *Y. angustissima*) are often abundant. In a number of places there are numerous areas of bare rock, acres in extent, devoid of plants of any kind except in the occasional crevices where a number of xeric or drought-resisting shrubs find a foothold. The dominant plants in the valleys are several kinds of low-statured wormwood (*Artemisia*).

Much of the Painted Desert is drained by the Little Colorado River, which rises in the pine-clad White Mountains of mideastern Arizona. The principal rocks of the region are brilliantly hued incarnadined sandstones and badlands of decomposed volcanic ash, often disposed in layers or bands of reds, oranges, yellows, and browns. Some of the soils made of volcanic ash absorb great quantities of moisture, and when wet, puff up to form what prospectors call self-rising soil. This not only gives a peculiar appearance to the landscape but also a most odd sensation to persons walking over it.

Perhaps the feature of greatest interest to the tourist in this desert is found in the Petrified Forest National Monument; there six separate fossil-log areas (called "forests") have been preserved for posterity. Most of the petrified trees now naturally survive in living form only in the Andes Mountains and the mountains in New Zealand. None of the fossil logs stand upright. The trees from which they were formed are estimated to be about 160 million years old and were evidently washed into a swampy basin, then later buried under layers of volcanic ash from near-by volcanoes, and finally supersaturated with silicon in solution. The old idea of molecular replacement of the woody tissue by silica has been abandoned in favor of a "supersaturation theory," which suggests that the wood material is still present. The brilliant coloring of the wood is due to a later straining by minute quantities of iron and manganese.

There are other national monuments within this area which should certainly be mentioned: these are Rainbow Bridge; Navaho National Monument, in which is the unique Betatakin Ruin; Sunset Crater; Walnut Canyon; and Wupatki. Each is well worth visiting in spite of the unimproved roads leading to them.

Each of the national monuments just mentioned maintains an excellent interpretive museum to inform the visitor of the particular interests in the surrounding area. For more comprehensive knowledge of the Painted Desert as a whole, one should see the well-planned and excellently made exhibits at the Museum of Northern Arizona, located just north of Flagstaff. This is a very active institution, carrying on continuing research on the natural history, geology, and ethnology of all of northern Arizona including the entire Painted Desert region. It interprets its findings to the public by means of its numerous graphic exhibits and by popular as well as many scientific publications.

Great credit should be given to this museum for instituting annual craft exhibits to encourage northern Arizona Indians in keeping alive their native skills in various lines from silversmithing to rug and garment weaving, pottery making, and other crafts. "The Hopi Craftsman" was started in 1930, the "Navaho Craftsman," in 1942. The exhibits are not only extensive but marvelously beautiful. The Hopi exhibits are on display at the museum for four days during the Fourth of July week end. The Navaho craftsman handiwork is shown during the last two weeks of July.

The flora and fauna of the Painted Desert show closest affinity to that of the Great Basin Sagebrush Desert; because of this some biologists believe the Painted Desert should be considered as just a southeastern extension of the Great Basin Desert.

Animal life is represented by many forms. Among amphibians can be found the familiar spade-foot toad (*Scaphiopus hammondii*), the red-spotted toad (*Bufo punctatus*), and the canyon tree toad (*Hyla arenicolor*). Lizards include the banded gecko (*Coleonyx variegatus*), collared lizard (*Crotaphytus collaris baileyi*, 42), leopard lizard (*Crotaphytus wislizeni*, 36), brown-shouldered uta (*Uta stansburiana*, 37), desert scaly lizard (*Sceloporus magister*, 41), sagebrush swift (*Sceloporus graciosus*), short-horned "horned toad" (*Phrynosoma douglasi*), whiptail lizard or race runner (*Cnemidophorus gularis*), and two less well-known species, the western earless lizard (*Holbrookia maculata*) and Stejneger blue-bellied lizard (*Sceloporus elongatus*). Of snakes there are relatively few: the striped racer (*Coluber taeniatus*), the gopher snake (*Pituophis catenifer*), and the prairie rattlesnake (*Crotalus confluentus nuntius*) that is used by the Hopi Indians of Arizona in their snake dances.

As might be expected the rodents are the most common mammals. The list of species includes the familiar antelope ground squirrel (*Ammospermophilus leucurus*, 118), Hopi chipmunk (*Eutamias quadrivittatus hopiensis*), Arizona wood mouse (*Peromyscus maniculatus rufinus*), long-nosed white-footed mouse (*Peromyscus nasutus*), wood rat (*Neotoma lepida* and *Neotoma mexicana*), Painted Desert kangaroo rat (*Dipodomys ordii longipes*), pocket mouse (*Perognathus apache*), grasshopper mouse (*Onychomys leucogaster pallescens*), Zuñi prairie dog (*Cynomys gunnisoni zuniensis*), Colorado rock squirrel (*Otospermophilus grammurus*), pocket gopher (*Thomomys bottae*); the rodent-like Texas jack rabbit (*Lepus californicus*), and the Colorado cottontail (*Sylvilagus auduboni warreni*) are also common over most of this colorful desert.

The larger carnivorous mammals include the ubiquitous coyote (*Canis estor*), badger (*Taxidea taxus,* 129), Arizona gray fox (*Urocyon cinereoargenteus*), ringtail cat (*Bassariscus astutus,* 124), bobcat (*Lynx rufus,* 128), and the spotted skunk (*Spilogale gracilis,* 133). Four common bats are found on the Painted Desert: the canyon bat (*Pipistrellus hesperus*), desert pallid bat (*Antrozous pallidus*), big brown bat (*Eptesicus fuscus*), and the long-eared bat (*Myotis evotis chrysonotus*).

The prong-horn antelope (*Antilocapra americana*) is not a true antelope. That it was once quite common over most of northern Arizona is evidenced by the widespread occurrence and great number of petroglyph drawings of this animal made by prehistoric Indians. At present this fleetest of all desert mammals is restricted to a few small isolated herds, all rigidly protected by law. In Arizona a few antelope can usually be seen from the highway between Holbrook and the south entrance to the Petrified Forest National Monument. Also quite often along this road can be observed, usually at night, a creature which seems quite out of place on the desert, the Arizona porcupine (*Erethizon epixanthum couesi*). It feeds on the bark of junipers, piñon, and cottonwood trees.

Common birds are: the red-tailed hawk (*Buteo borealis,* 68), prairie falcon (*Falco mexicanus,* 66), desert sparrow hawk (*Falco sparverius,* 67), mourning dove (*Zenaidura macroura,* 93), horned owl (*Bubo virginianus,* 56), Nuttall poorwill (*Phalaenoptilus nuttalli,* 83), white-throated swift (*Aeronautes saxatalis,* 80), black-chinned hummingbird (*Archilochus alexandri,* 96), red-shafted flicker (*Colaptes cafer*), Cassin kingbird (*Tyrannus vociferans*), ash-throated flycatcher (*Miarchus cinerascens,* 64), Say phoebe (*Sayornis saya*), violet-green swallow (*Tachycineta thalassina*), raven (*Corvus corax*), piñon jay (*Cyanocephalus cyanocephalus,* 74), chickadee (*Penthestes gambeli*), rock wren (*Salpinctes obsoletus,* 102), gnatcatcher (*Polioptila caerula*), house finch or linnet (*Carpodacus mexicanus*), and the desert sparrow (*Amphispiza bilineata deserticola,* 104).

PLATES

DESERT *Insects*

In spite of the severe conditions of life the desert's insect population is seasonally large. The grasshoppers and their relatives (*Orthoptera*) are said to be the most common, then follow the butterflies and moths, particularly the latter; beetles, wasps, bees, and ants come next in numbers. Only during the colder months are insects not much in evidence. As soon as the warmer days come and the wild flowers begin to bloom, insects of many kinds, sizes, and colors are seen. One has then only to light a lantern or sit around the campfire at night to realize how many insects have adapted themselves to live in this land of drought. They come into the light from every direction and often in unbelievable numbers; especially abundant are the small moths.

Many times in spring I have seen the ground almost black with large red-headed, black-winged blister beetles feeding on evening primroses. As they awkwardly crawled over the small rocks or were quietly eating there could be heard a strange rasping noise made by hard chitin-covered bodies scraping on the rocks or by the striking together of myriads of strong hard mandibles as the insects chewed the juicy or leathery plants. Often I have seen too, in the spring season, a colorful living, moving carpet of creeping sphinx moth larvae, devouring almost every green herbaceous plant before them. Crickets at times come in plagues in the summer to harass the desert's human residents and visitors. Near irrigated areas, but sometimes many miles away, mosquitoes may annoy at night. In the Sonoran deserts a large heavy-bodied, slow-flying mosquito belonging to the genus *Theobaldia* often disturbs one's slumber by its deep fly-like buzz.

Some of the desert insects, particularly the small bees, have a peculiar way of avoiding the perils of prolonged severe drought. The time of emergence of the young bees from their underground cells is uniquely synchronized with the blooming of the particular flowers on which they will feed. The same moisture which stimulates the flower's growth serves to initiate the development of the young bees. If no favorable rains fall the immature pupa may hold up development into the mature stage for several years.

It is well known that the eggs of certain desert insects tolerate great dryness, even actual loss of water by the embryo, and still remain viable over long periods, even up to several years.

It is amazing how far away from water one can observe dragonflies. These great strong-winged, keen-eyed insect hunters fly far and wide in the most arid places. It is only at the time of egg laying and during the nymphal stages that they need to have water. Most of the desert species are of medium size and have somber-colored wings and body. In the near vicinity of fresh-water ponds, and lakes and rivers, many of the larger, colorful species common to western United States and Mexico are seen. One of the smallest of all the damsel flies, *Anomaloagrion hastatum,* is found in the Colorado Desert along the Colorado River, although not limited there. It is a most delicate insect of plain gray color and with a wing spread of but 21 mm. (¾ inch)! It is most commonly seen in the bright warm days of summer.

Several species of the much-specialized small ephydrid flies are found around alkaline ponds and saline lakes of the mid-desert. The active cylindric larvae of *Ephydra gracilis* live suspended in the very salty waters of the Salton Sea and Great Salt Lake. The adult flies often swarm in enormous numbers above the surface of the water and along the shores. The Paiute Indians formerly collected and dried the pupae which sometimes were cast up in elongated heaps or windrows on the shores after windstorms lashed the waters of the lake. The food they prepared from them was called Koo-chah-bie.

Only a few of the more commonly seen insects and other arthropods are here illustrated and described but to know even these few will make the desert's arthropod population enormously more interesting and worthy of study.

1 PRAYING MANTIS. Family *Mantidea*. A representative of a group of peculiar, largely tree-dwelling insects in which the forelegs are held in front of the face as if in prayer. They appear to be wholly carnivorous, devouring their prey alive. Pugnacious and, we may say, cruel. The peculiar egg capsules are attached to plant stems.

2 ANT-LION LARVA or DOODLE BUG. Family *Myrmeleonidae*. Called ant-lions because of their habit of devouring ants which they seize while lying in ambush at the bottom of their funnel-shaped pits. These lurking-holes are usually made in fine sand or dust on the protected sides of stones or clay banks, or under dense shade of shrubs.

3 ADULT ANT-LION. *Scotoleon longipalpus*. The gray-winged, soft-bodied adult ant-lions frequently come to the presence of lights. They usually begin to fly at dusk. They are sometimes mistaken for damsel flies.

4 WALKING STICK. *Parabacillus hesperus*. This peculiar orthopterous stick-like insect represents a group of walking sticks which are wingless. They feed on the leaves of trees and shrubs. The small unprotected seed-like eggs are dropped among dead leaves lying on the ground where they hatch in the following spring. The related species, *arizonicus*, occurs in Arizona, adjacent Mexico, and southern California.

5 SILVER-SPOTTED GRASSHOPPER. *Bootettix punctatus*. This is one of the slant-faced locusts that live on creosote bushes. The body is green with brown, black, and silvery-white markings. So closely do the insects match the creosote bushes that it is difficult to locate them. They make a distinctive wispy, clicking sound, especially noticeable about sunset when the wind goes down.

6 MEXICAN GROUND CRICKET. *Nemobius mexicanus*. This represents a large group of uniform-brown, often very abundant crickets of arid portions of western North America. In Arizona and in the Imperial and Coachella valleys in California they often constitute a summer pest, getting into houses and stores in annoying numbers. They are sometimes migratory.

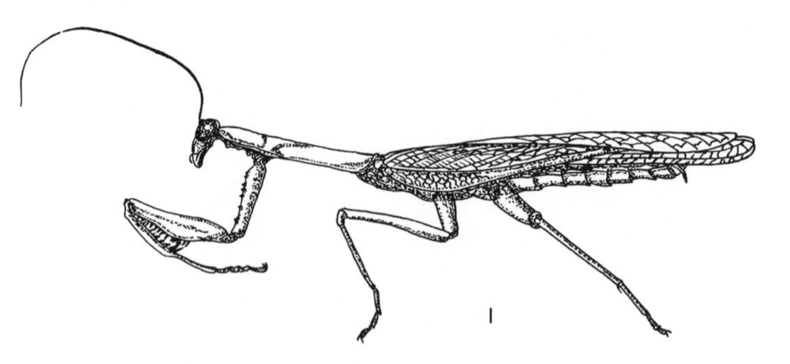

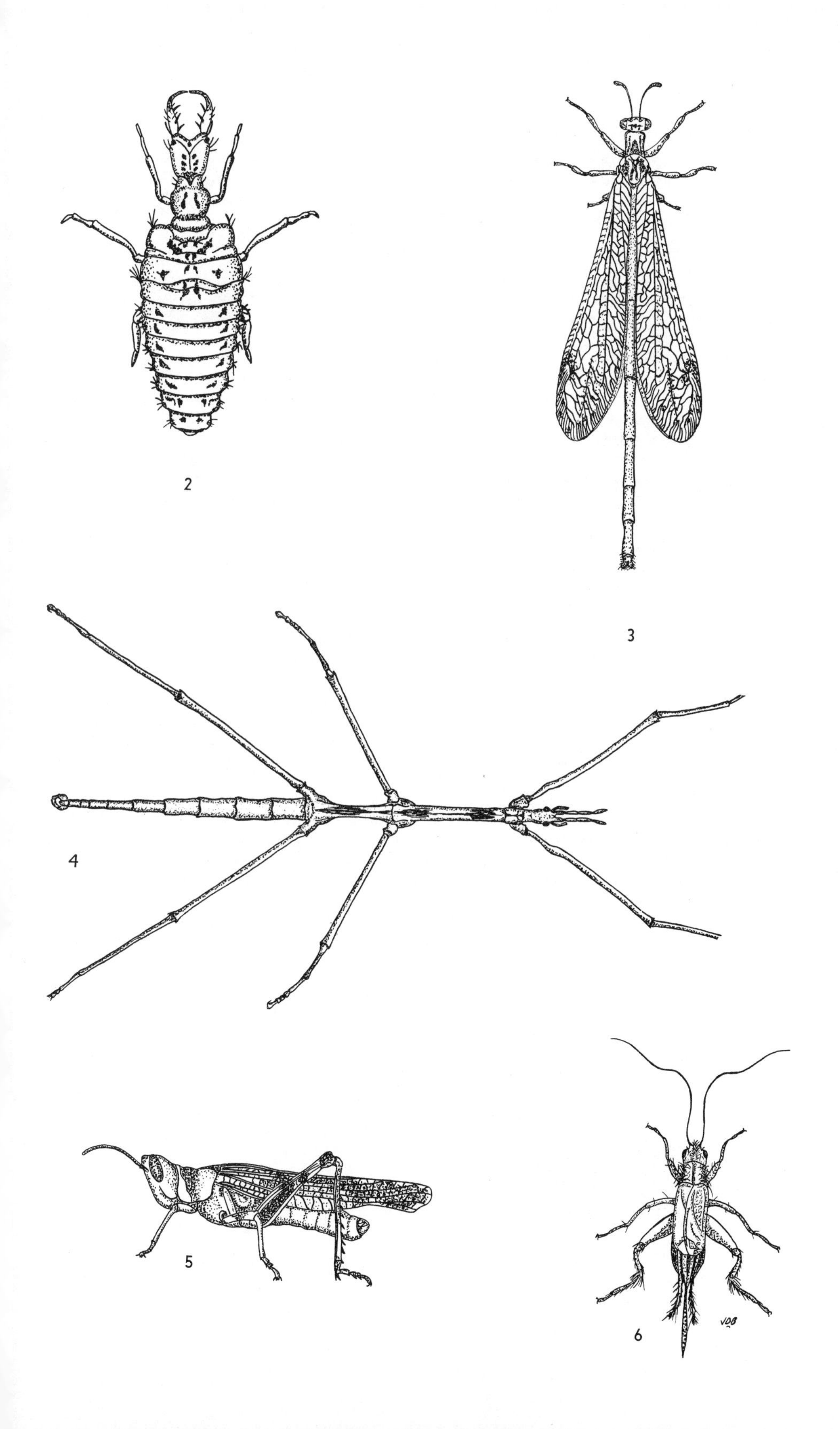
2
3
4
5
6
VDB

7 MESQUITE BORER. *Megacyllene attenuatus.* The brown round-headed mesquite borer has a large wood-boring, brown-headed larva with cream-colored body. The strong jaws work rapidly through recently cut or cured mesquite fence posts and logs, soon reducing them to yellowish powder. The adult beetle may be 1½ inches long.

8 WESTERN CONE-NOSE BUG. *Triatoma protracta.* A representative of blood-sucking bugs found in pack rat nests and capable of transmitting a form of sleeping sickness. Other somewhat similar-appearing bugs feed on the sap of plants.

9 GRAY-SNOUT WEEVIL. *Eupagoderes desertus.* One of the commonest beetles found on herbs and shrubs. Its gray body enables it to escape detection until it moves or until it drops to the ground while simulating death. The long-snouted small head fits into the thorax, like the ball of a ball-and-socket joint.

10 IRON-CLAD BEETLE. *Phloedes pustulosus.* These slow, death-feigning, black, rugose beetles with unbelievably tough, hard, and thick exoskeletons are sometimes found on the ground. They normally live under the bark of dead trees and tree stumps. When feigning death, after being disturbed, the antennae and legs are fitted into special grooves of the hard body, for protection. One must use a hammer to drive a pin through the body before mounting the specimen in one's insect collection.

11 CIRCUS BEETLE. *Eleodes armata.* Called circus bugs or stink beetles because of their queer way, when disturbed, of standing with the body raised vertically and because when handled or crushed they give off a strong offensive odor. Single individuals often come walking slowly into the camp and are considered conveyors of good luck. Sometimes in autumn they collect in large numbers. They feed on dead and decaying vegetable matter. The body is jet black and oily looking.

12 SOLDIER BEETLE. *Lytta magister.* This is a representative of the soldier or blister beetles. The transparent blood when applied to the skin causes blisters, hence the name. Blister beetles of the related genus *Tegrodera* are extraordinary desert species with red heads, shining black bodies, and wings covered with small indentations in which pollen collects as they feed on flowers. They may occur in extraordinary numbers in spring while crawling over the ground as they feed and mate.

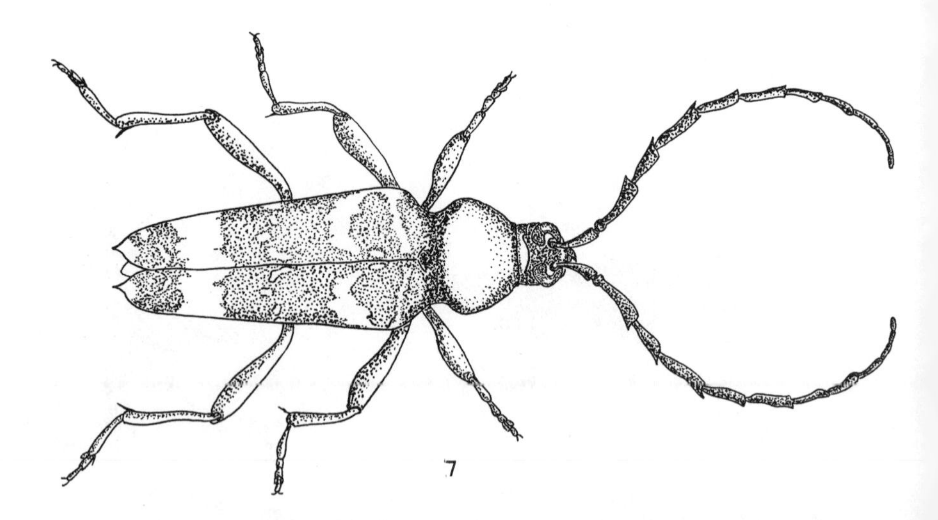

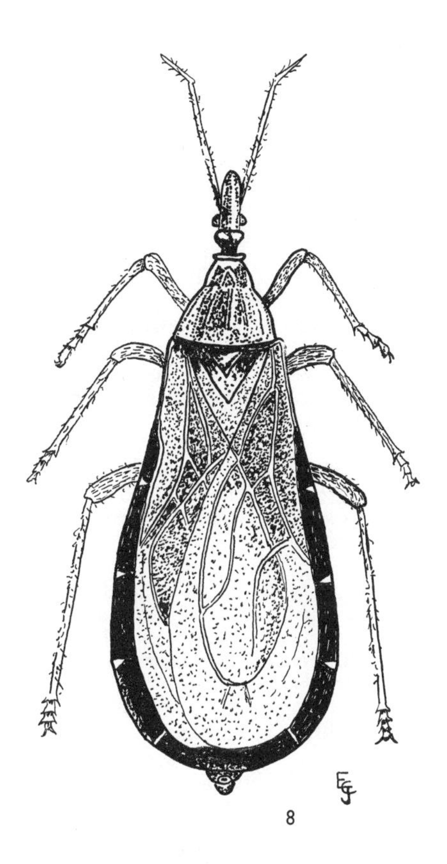
8

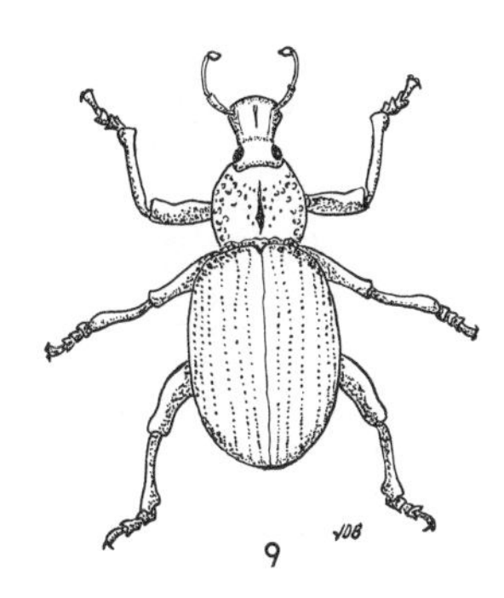
9

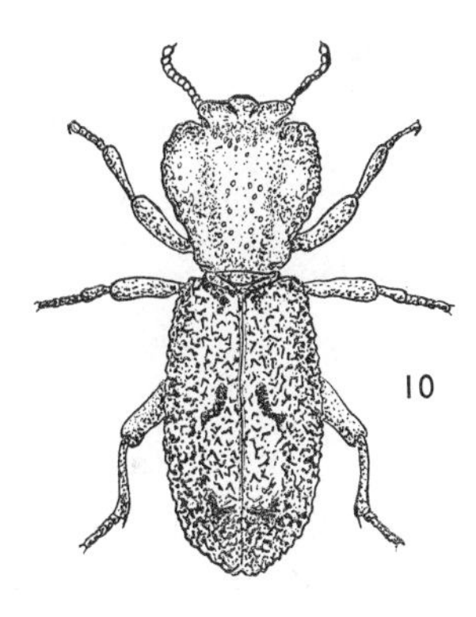
10

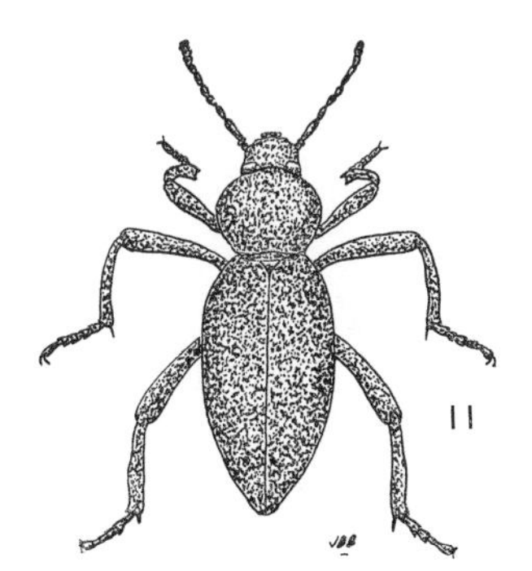
11

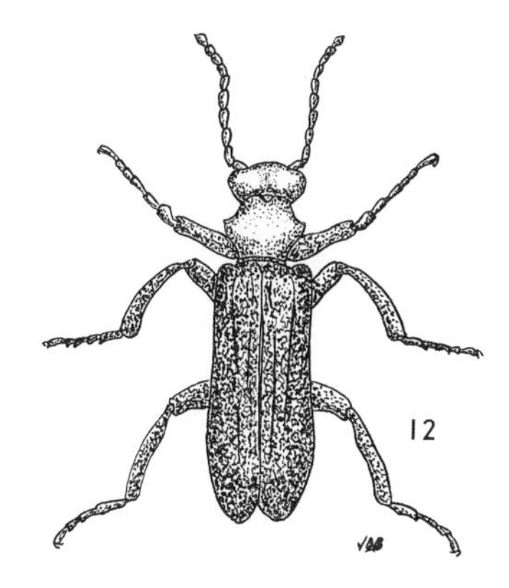
12

13 PAINTED LADY. *Vanessa cardui.* Of world-wide distribution, this butterfly with red, black, white, brown, and blue markings often occurs in enormous numbers in desert areas, sometimes in great migratory flights. There may be two broods in the year. The caterpillars feed on a variety of plants.

14 BECKER'S PIERID. *Pieris beckeri.* This fine butterfly feeds on the yellow flowers on the bladderpod (*Isomeris*), which blooms throughout many months of the year and occurs on the Colorado and Mohave deserts and on the deserts of western Sonora.

15 SPHINX MOTH. *Celerio linata.* This handsome swift-flying sphinx moth is often found in great numbers in the Sonoran Desert. The black, green, and yellow larvae feed greedily on the sand verbena, various species of evening primroses, and other desert plants, and often occur in unbelievable numbers in March and April and constitute a pest. The fat caterpillars were used as food by the Cahuilla Indians who fried them on hot earth from which the coals of creosote bush fires had been scraped.

16 INFLATED BEETLE. *Cystodemus armatus.* This blackish beetle with arched wing-covers may be seen bustling about on the sand or on flowers where it feeds. The pits of the wing covers are often filled with loose pollens so that they appear yellow or cream colored.

17 CALIFORNIA PALM BORER. *Dinapate wrightii.* This large shiny, dark-brown beetle, once considered so rare that specimens sold to museums for several hundred dollars apiece, is now known as a frequent inhabitant of desert areas where Washingtonia fan palms occur. Its larvae live on the wood of dying and dead palms, and there the beetles pupate. Holes of large size in old palm logs indicate the former living quarters of the immature developing insects. Colorado Desert in California, and Baja California.

18 YUCCA BORER. *Scyphophorus yuccae.* These black, long, curved-snouted beetles are found in spring crawling among the crowns of leaves of yuccas. The eggs are laid there and the larvae bore into the tender bud tissue and feed. Later they pupate in the stem and form tough small peanut-size cases of fibers. They may kill the upper end of the branch.

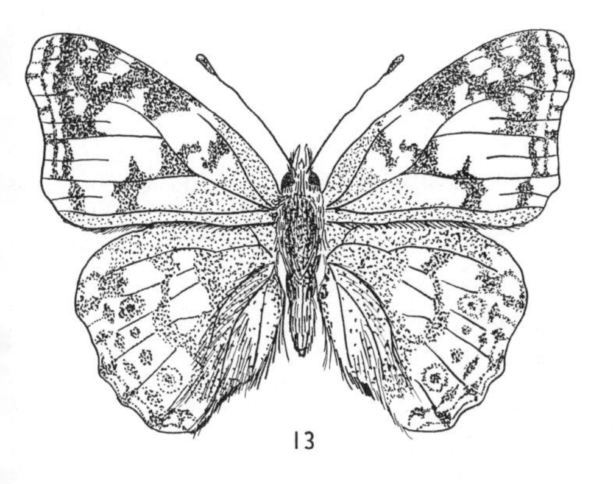
13

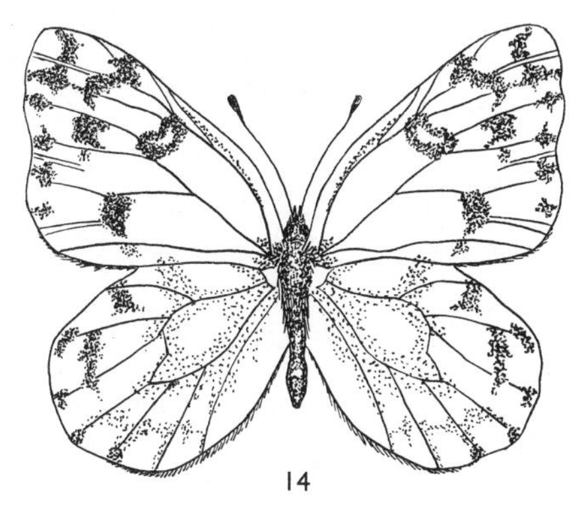
14

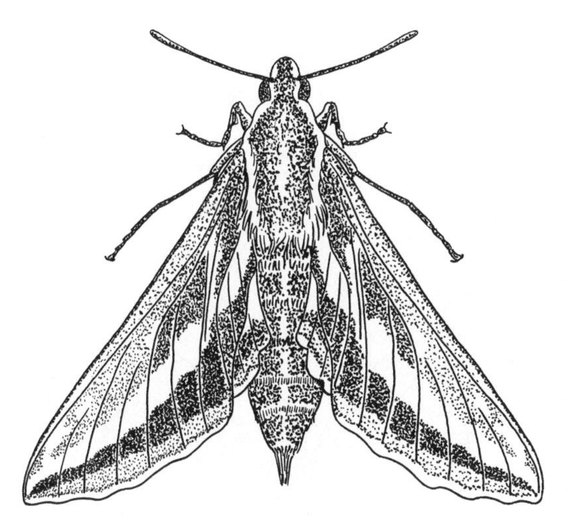
15

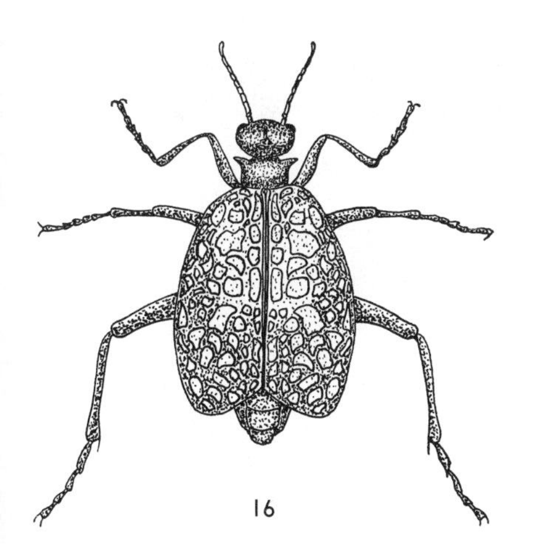
16

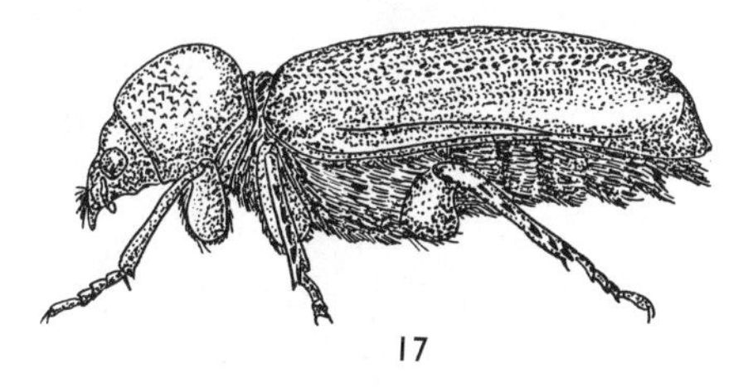
17

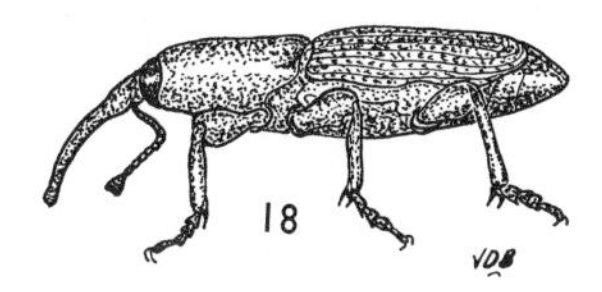
18

19 TWO-TAILED SWALLOWTAIL. *Papilio daunus.* This large yellow, two-tailed papilio is confined to the northern Mohave Desert. Other "swallowtails" are found in desert areas feeding on flowers of the numerous kinds of milkweeds.

20 RUFESCENT PATCH. *Chlosyne lacinia crocale.* This striking black butterfly, local in California on the Colorado Desert, is widely distributed in Arizona, New Mexico, and Texas. It feeds on the numerous species of sunflowers (*Helianthus*).

21 MILKWEED BUTTERFLY. *Danaus menippe.* This widespread and common butterfly has beautiful orange-brown wings bordered and striped with black, and many white spots in the black areas of the wings. They migrate northward in summer and southward in winter. A distinctive dark spot occurs near the middle of the lighter-colored hind wing of the male.

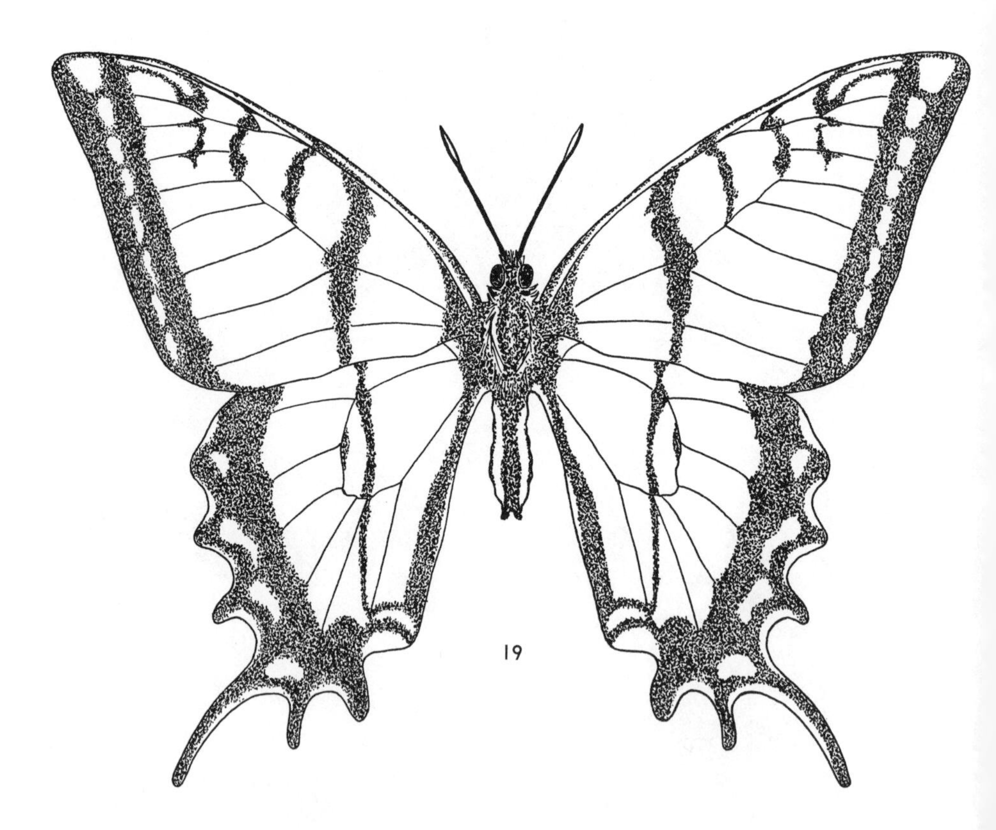

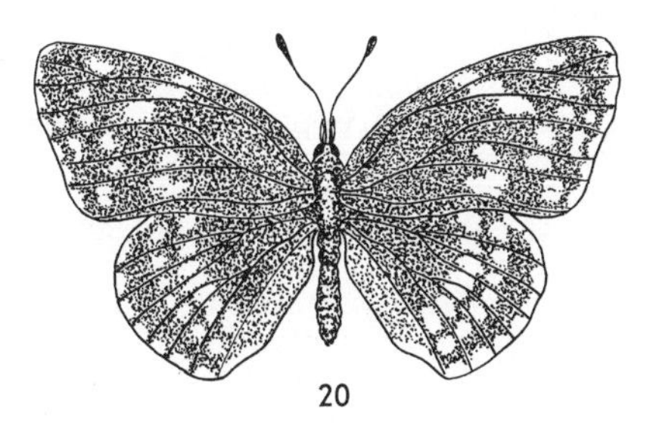

20

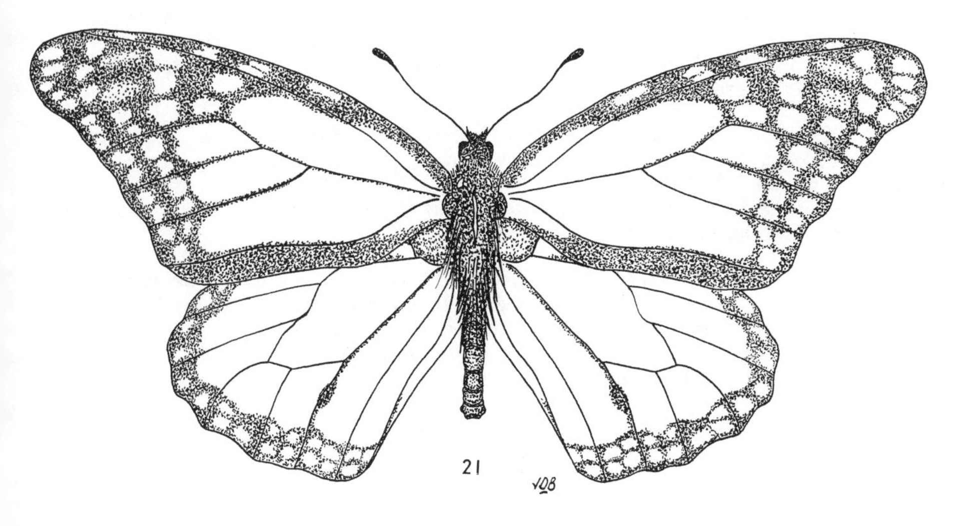

21

22 TABANID or HORSEFLY. *Tabanus punctifer.* This large black fly is not only a pest on cattle, horses, and deer but on human beings too. They are most common during hot summer days. The larvae live in mud or shallow water of ditches and ponds. Widespread on deserts.

23 MEGANDRENID or BURROWING BEE. *Megadrena enceliae.* The megandrenids are solitary or social bees constructing their burrows in the ground or in clay walls of canyons. The burrows are provisioned with nectar and pollen. Both males and females remain in the nests throughout the fall and winter and emerge in the spring.

24 MUTILLID WASP. *Dasymutilla satanus.* Children generally call these hairy, solitary, wingless wasps "hairy ants" or "fuzzy ants." Some species have long brick-red hairs covering the upper body parts, while others have an adornment of white or yellowish hairs. Only the males have wings. These are very tough little insects; they are fast crawlers and bear a long sting.

25 CARPENTER BEE. *Xylocarpa californica.* This is a representative of a group of robust bees which cut tunnels in dry wood where they nest. In the desert area they often choose cottonwood trees in which to work. The cells are provisioned with a mixture of honey and pollen. The large females are generally metallic black, dark blue, or green. The males, often smaller, have yellow to whitish hairs forming a velvet on the thorax and abdomen.

26 PALM or APACHE CICADA. *Diceroprocta apache.* This is one of the shrill, noise-making insects of the Sonoran deserts. It lays its eggs in the various species of palms, including the date palm of commerce. It is preyed upon by the cicada killer.

27 CICADA KILLER. *Sphecius convallis.* One of the largest American wasps. It has black and yellow bands on the body. In summer it can be seen carrying paralyzed cicadas with which it provisions its nest. Resident of desert areas where palms grow.

28 TARANTULA HAWK. *Pepsis thisbe.* This beautiful large wasp feeds on flowers. The wings are various shades of blue-violet, orange, or red. The blue-black body is smooth and metallic. The nest is provisioned with various spiders, including trap-door spiders and tarantulas; these furnish food for the developing larvae.

29 ARIZONA TERMITE. *Gnathamitermes perplexans.* One of the many wood-eating termites of the desert areas. It is a pale brown species; the soldiers are pale yellow. On the stems and leaves of agave, yucca, and other shrubs the termites construct a mud coating under which they work. Most readily seen after summer rains.

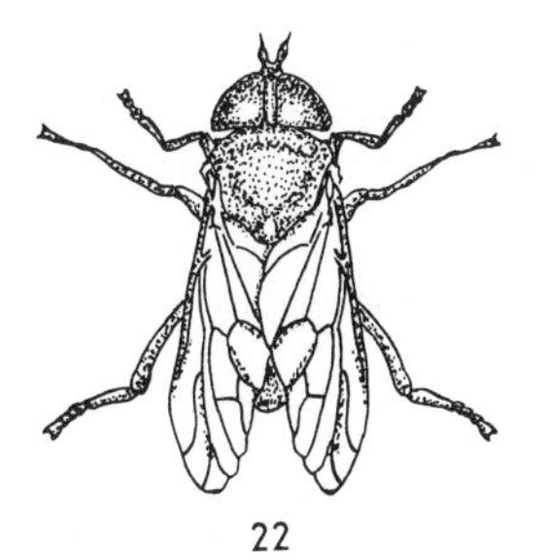

22

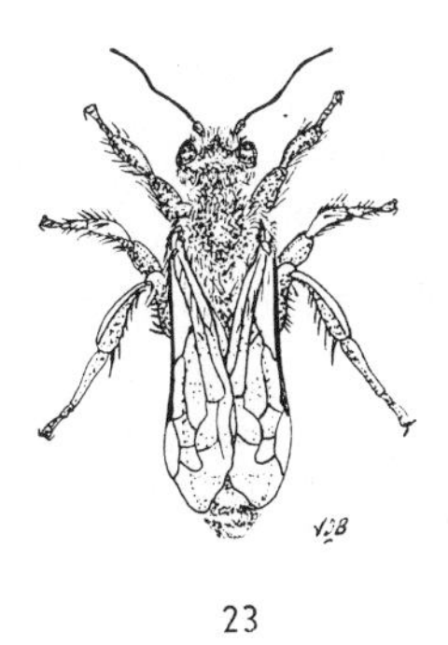

23

24

25

26

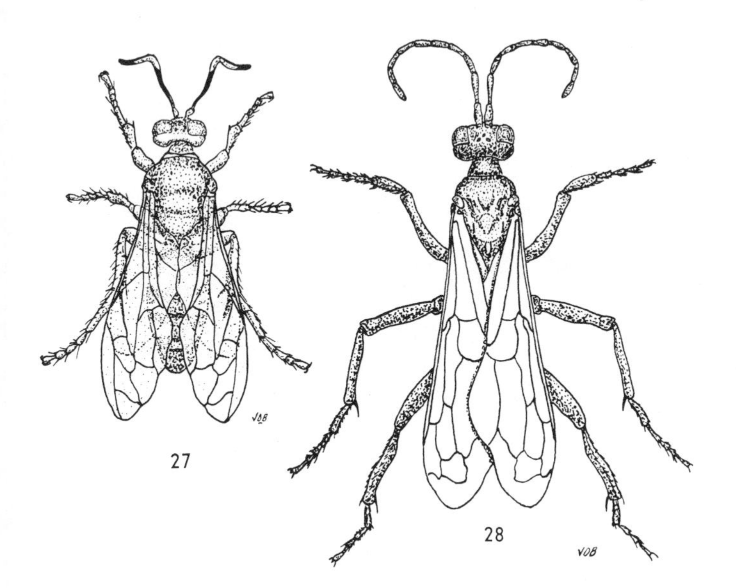

27

28

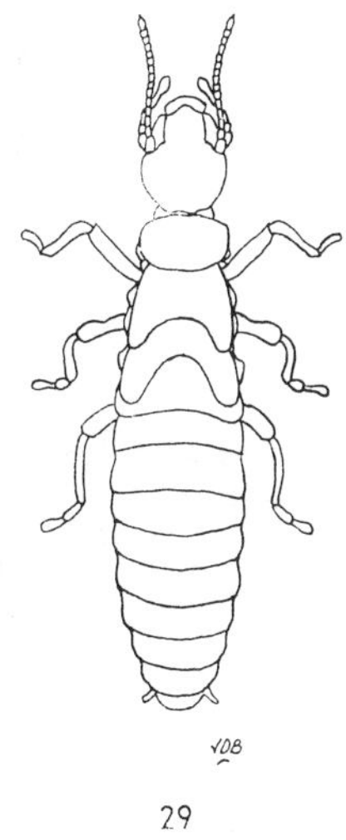

29

30 CENTIPEDE. Centipedes are largely night crawlers. Specialized appendages ending in sharp perforated claws are found beside the head; these can inject a poison, but centipedes are not particularly dangerous to man. Appendages at the end of the abdomen are used as tactile organs. Centipedes are voracious feeders and speedily cut up and devour insects they capture. Widespread in deserts. There is one pair of appendages to each abdominal segment.

31 MILLIPEDE. These elongate, round-bodied creatures vary in color from brown to almost black. Each of the many segments bears two pairs of walking appendages. Most kinds of millipedes are harmless but a few are irritating to the skin. The food is vegetable matter. When handled they coil up and often emit a foul-smelling fluid which stains the hands brown. The desert millipedes are small compared with most species living in humid areas. They burrow into sand for protection and to get moisture.

32 SOLPUGID. These are light tan, hairy, "spidery looking" arthropods which generally are seen crawling rapidly over the ground at night in the light of the campfire, or on screen doors of lighted houses where they are hunting their insect prey. They are perfectly harmless.

33 WHIP SCORPION. These harmless but "horrible looking" near-relatives of spiders somewhat resemble true scorpions but, instead of having a jointed tail ending in a poisonous sting, have only a long but stout whiplike bristle terminating the abdomen. They are largely nocturnal and hide under stones and fallen wood by day. The body is dark brown in color. Absent from far western and northern desert areas. The food is chiefly insects.

34 SCORPION. These arthropods range in size from 1¾ to 4½ inches, when full-grown. The largest kinds are generally comparatively harmless creatures but some of the smaller ones of southern Arizona and Mexico have a very poisonous sting. They are very active on warm nights when they crawl about looking for insect prey. Persons sleeping on the ground in the warmer months of the year may find them crawling under their covers. The sting of any scorpion is painful.

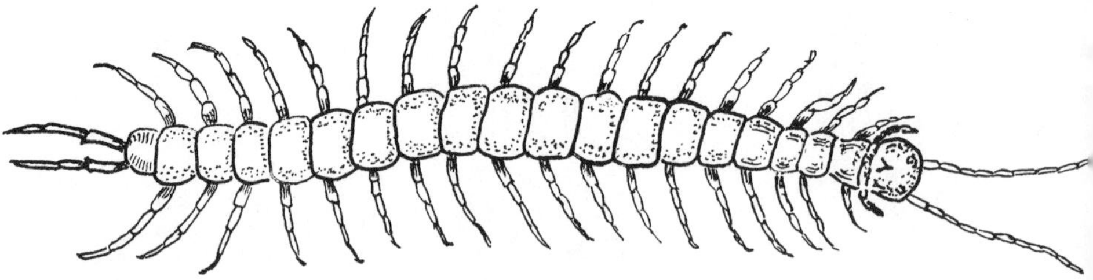

30

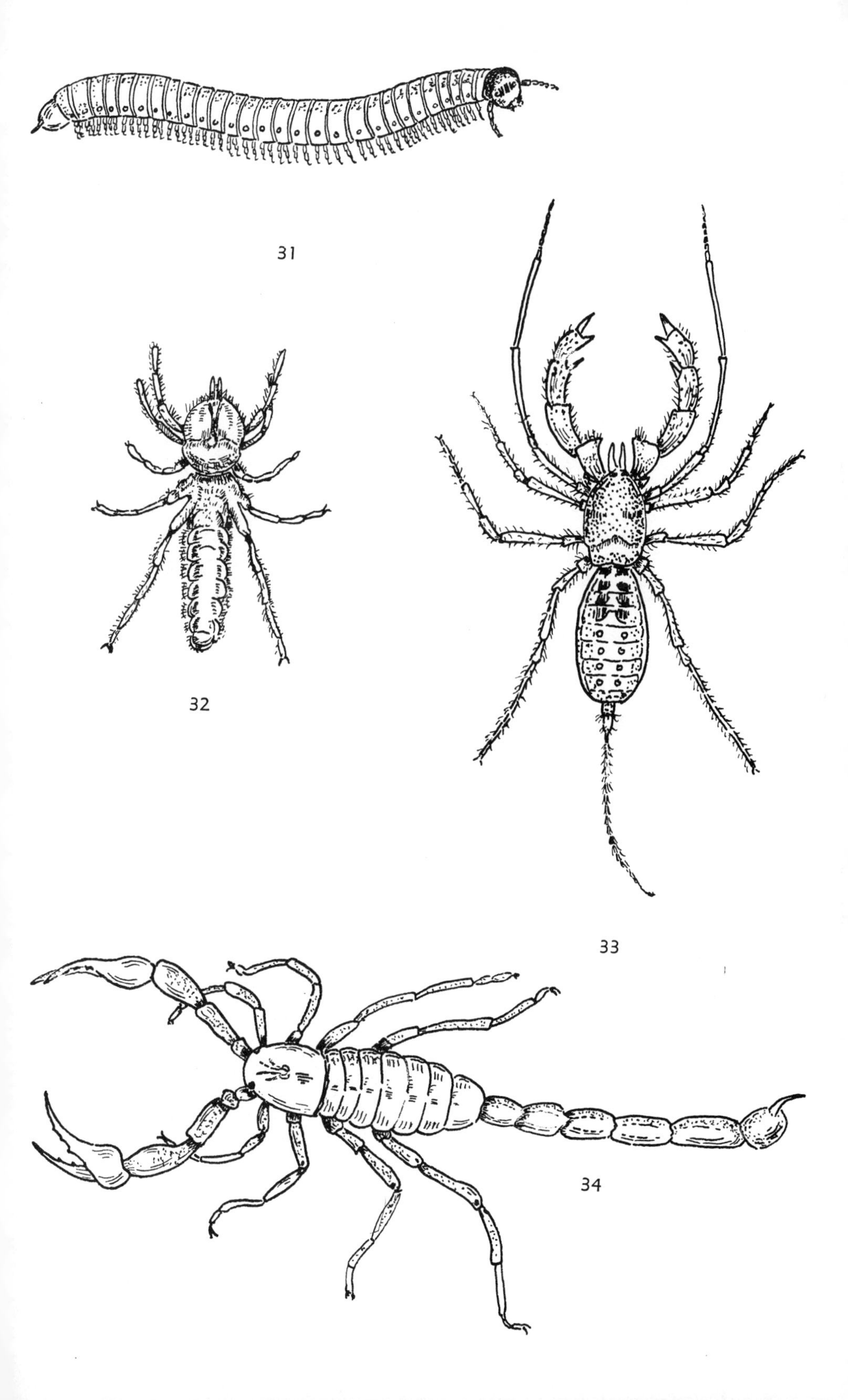

31
32
33
34

DESERT *Reptiles*

Reptiles are among the commonest, most conspicuous animals seen on the desert—the lizards by day, the comparatively few snakes by night, especially during the hotter months of the year. The snakes do most of their hunting during hours of darkness or semi-light and this accounts for their activity then. The direct heat and dazzling light of the summer sun, snakes usually avoid; to some of them direct prolonged exposure to the sun is fatal. The lizards seem to rejoice in both diurnal heat and light, and sometimes we are much surprised to see them in the spring season basking or moving about stalking prey while on the surface of sands which glow with tremulous light.

However, even the hardy lizards have their limits of heat tolerance. The normal favorable temperature for greatest activity for many kinds is near 100° F.; when the temperature mounts to 110° and upwards to 117° F. they are usually found hidden in the shade of bushes, lying buried in the sand, or seeking refuge in the burrows of rodents. The swift gridiron-tailed lizard (*Callisaurus draconoides,* 35) appears to be "the demon supreme" for high-temperature tolerance. Snakes are known to be most active and efficient as hunters between the approximate temperatures of 68° and 80° F. When the temperature goes above 98° and up to 107° F. most of them lose co-ordination and go into a state of heat prostration. Most lizards can take a temperature a little higher before going into a state of heat rigor. In the so-called "winter months," November, December, and January, about the only lizard seen is the harmless, agile, and alert little side-blotched lizard (*Uta stansburiana*); the snakes, along with most of the lizards, are then hidden beneath sands and rocks and lie in a state of hibernation stupor.

Many of the desert lizards are fierce carnivores or constant cannibals, preying on other lizards, sometimes on individuals larger than themselves. Their food is not chewed but swallowed whole while still alive. Other kinds live mostly on insects such as ants, termites, flies, and caterpillars; still others, like the chuckawalla, are vegetarians, eating mostly flower petals and tender plant shoots and leaves.

Neither lizards nor snakes are regular water drinkers but depend mostly on the fluids found in the plants or in the bodies of the animals they eat. This is to them a very great advantage in a waterless environment.

Dr. Robert C. Stebbins of the Museum of Vertebrate Zoology of the University of California has shown that certain of the sand-burrowing lizards have special adaptations to keep out, or at least capture, fine soil particles before they reach the lungs. In most of the lizards there is erectile tissue near the entrance of the nose and also near the pathway of entry to the moist inner nasal chamber. This tissue may be elevated so as to reduce the size of the nasal canal. Beside these adaptations there may be a large secreting gland whose flow of mucus moves outward any dust or other foreign particles that get to the principal nasal chamber. These provisions of nature are very important to inhabitants of deserts, where dust and sand are common.

Interesting similarities in structure of distantly related lizards, snakes, and other animals in widely separated deserts of the world are often cited as examples of convergent evolution due to the impress of similar environments. Thus, the rock-dwelling American desert chuckawallas (*Sauromalus*) are matched in the Sahara by the spiny-tailed lizard (*Uromastix*). Their habits are very similar, even to their vegetarian diet. The only readily recognized outward difference is in the larger, heavier tail scalation of the Saharan lizard. The horned lizards (*Phrynosoma*) of the sandy and rocky American deserts have their parallel in the strange flat-bodied, spine-covered moloch (*Moloch horridus*) of the Australian midcontinent deserts.

Contrary to popular belief the desert wilderness is not a place teeming with "crawling, creeping reptiles." Snakes, lizards, and tortoises there are, to be sure, but not many. The lizards, all wholly harmless except the Gila monster, are the most numerous. Persons who adopt habits of ordinary caution have little to fear even from the venomous snakes. Unless particularly looking for them one may see but few in a whole season of warm days. It is not advisable to walk abroad on the desert on warm summer nights without carrying a flashlight or lantern. Snakes then are readily detected. Such a practice may save one the horror of getting too near or stepping upon a dangerous rattlesnake.

Of the many kinds of reptiles to be seen in desert places only a few of the commonly met or most unusual ones are illustrated in the following pages. But even to know these is a good beginning toward fuller knowledge of these cold-blooded desert denizens. To those interested in exact identification and a knowledge of the habits of arid-land reptiles I commend *Amphibians and Reptiles of Western North America*, by Robert C. Stebbins.

35 ZEBRA-TAILED LIZARD. *Callisaurus draconoides.* A very light-colored, smooth-scaled lizard simulating the whitish sands and gravels upon which it runs with great speed. In running it shows the black and white markings of the tail, which it carries curled forward. It is largely an insect feeder but may eat other lizards. Deserts of southwestern United States from central Nevada to Mohave, Colorado, and Yuman deserts, thence southward into Baja California and Sonora.

36 LEOPARD LIZARD. *Crotaphytus wislizeni.* 12–16 inches. Gray above, white below; back and upper surface of limbs covered with dark brown spots; back and tail crossed by narrow light bars; throat white. In breeding season the female has bright red or salmon-colored vertical streaks on sides, an unusual condition since among most lizards it is the male which is the more brilliantly colored. Diurnal. Food: insects, other lizards, some vegetation. Range: sandy open portions of all North American deserts.

37 BROWN-SHOULDERED or SIDE-BLOTCHED LIZARD. *Uta stansburiana.* A small, very active lizard widely distributed in arid areas from below sea level to as high as 7,000 feet in the mountains. Often it is the most abundant lizard found, and in warmer areas it is active all winter. Called side-blotched because of the usual black spot behind the "arm pits."

38 FRINGE-TOED or OCELLATED SAND LIZARD. *Uma notata.* 6–9 inches. Light gray overlaid with a network of black lines through which the ground color appears as light islands, each with a central red-brown spot; belly almost white, marked on either side with a black blotch edged with red-orange. Hind toes fringed with elongated scales to aid in running over sandy surfaces and in digging. Special nasal valves and shovel-nosed snout to permit plunging into and "swimming" in dune sand. Diurnal. Food: small insects. Range: sandy areas of the Colorado and Yuman deserts. A closely related species, the Coachella uma (*Uma inornata*), is restricted to the northern portion of the Colorado Desert.

39 DESERT BANDED GECKO. *Coleonyx variegatus.* 4–6 inches. Pale yellowish-gray; head and back spotted or banded with red-brown; resembles a salamander in appearance; almost translucent. Tail is storehouse for fat, its size indicating the amount of food reserve. Nocturnal. It has enemies in several night-crawling snakes. Food: small insects. When annoyed, it makes a high-pitched series of protesting squeaks, the only known instance of a North American desert lizard making a sound other than a hissing. Range: under rocks and dead limbs or in rodent holes of the Mohave and Sonoran deserts. A related species, the TUBERCULAR GECKO (*Phyllodactylus tuberculosus*), with leaf-like toes, occurs from Baja California north to the Southern Colorado Desert.

35
36
37
38
39

40 GILA MONSTER. *Heloderma suspectum.* This grotesque saurian is named after the Gila River near which it is found (*gila* appears to be a word of Indian origin meaning "spider" [Barnes, Will C.]). There are two species of Gila monsters; one (*H. horridum*) known as the MEXICAN BEADED LIZARD, found along the Mexican west coast as far north as southern Sonora; the other (*H. suspectum*), found over much of desert Sonora, enters southern Arizona and the most southwestern parts of New Mexico. It is a Lower Sonoran species of mesquite and creosote bush desert. Although poisonous, it is protected by law in Arizona. The Gila monster is a rather sluggish, harmless creature unless annoyed.

41 SPINY or ROUGH-SCALED LIZARD. *Sceloporus magister.* A robust, wary lizard with many pointed scales. It lives on the ground or, where yuccas abound, it runs over the stems or hides beneath the spines; often, too, it is seen about pack rat "nests." The females may lay up to 18 eggs. The male may show a blue throat-patch.

42 COLLARED LIZARD. *Crotaphytus collaris baileyi.* 12–14 inches. Brown or tan color crossed on back by several white or yellow lines and sprinkled with light spots; neck crossed by two black bands separated by white. Diurnal. Food: lizards, large insects; most vicious of desert lizards and has a voracious appetite. When in flight, may run on hind legs with long tail used as balancer. Range: normally, boulder-strewn areas over lower parts of all six deserts.

43 AGASSIZ TORTOISE or DESERT TORTOISE. *Gopherus agassizi.* A land tortoise with high-arched dull-brown back, reaching up to 13 inches in length. Shields show well-pronounced growth rings. Limbs are stocky and without sign of web. Nails are blunt except in the young. Males with concave lower shell (plastron) and with heavy gular shield; tail of males longer than that of female. Southern Mohave Desert, southeastern California, western Arizona, and most of Sonora. Fossil specimens have been found in the Pleistocene McKittrick tar beds of the southern San Joaquin Valley.

44 DESERT HORNED LIZARD or DESERT HORNED TOAD. *Phrynosoma, sp.* The specimen illustrated is the Texas horned lizard (*Phrynosoma cornutum*), with 2 rows of abdominal fringe scales. Horned lizards are found on all the southern deserts. On the Great Basin, southern Mohave, and Colorado deserts occurs the species *platyrhinos* with short head-spines and blunt snout; on the sagebrush plains of Utah is the species *douglassi* with very short head-spines; in the middle of Arizona and south into Sonora occurs the species *solare* known as the REGAL HORNED LIZARD. The flat-tailed horned lizard of the species *m'calli* is confined to the Colorado and Yuman deserts.

45 WESTERN WHIP-TAIL LIZARD. *Cnemidophorus tigris.* Known for its striking gait of "short nervous jerks." It is a swift runner; at top speed its long tail is lifted slightly above ground. Insects, scorpions, and spiders are its chief food. Great Basin, Colorado, and Mohave deserts eastward and southward into Arizona, New Mexico, Texas, and Sonora and the Central Plateau of Mexico.

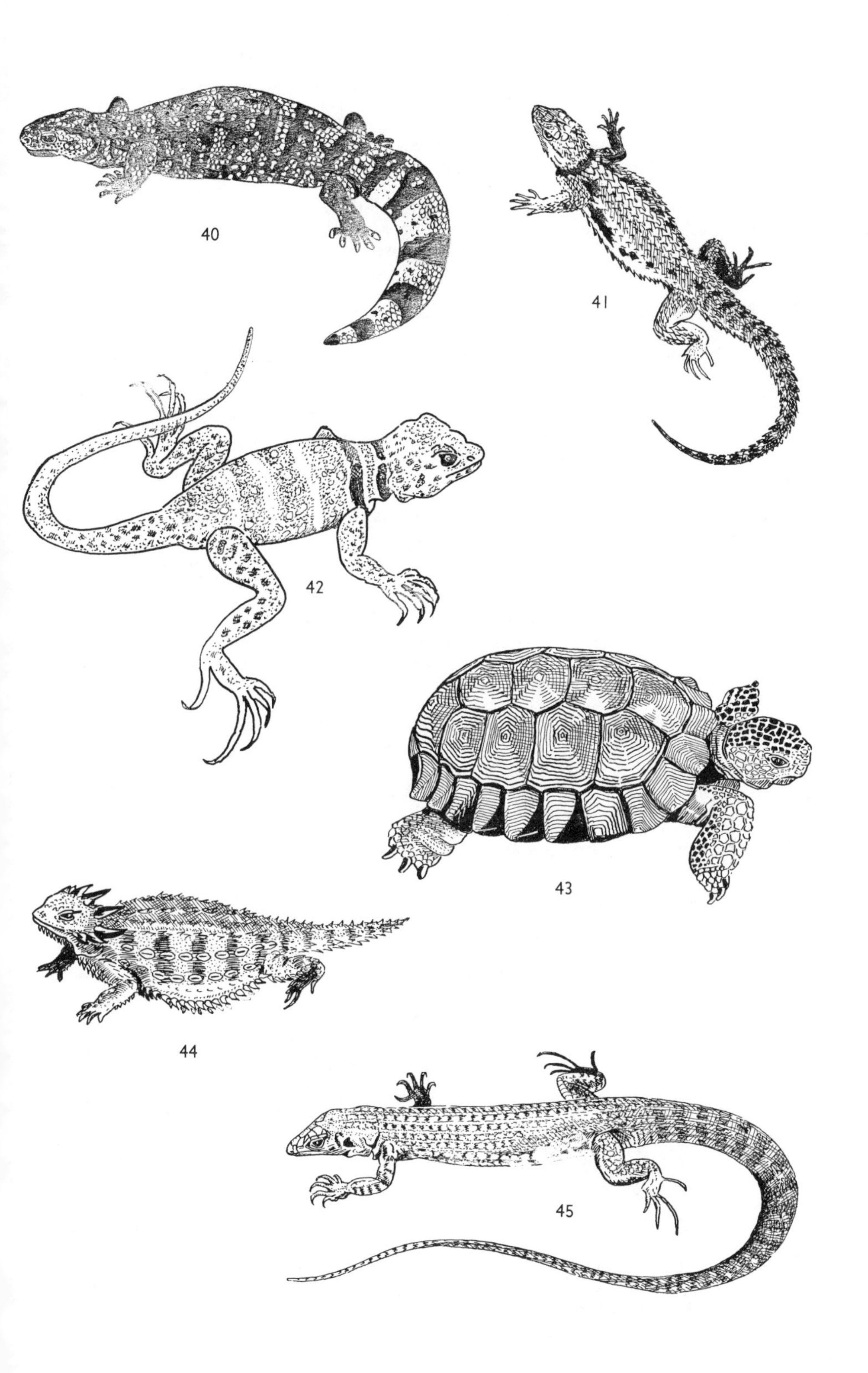
40
41
42
43
44
45

46 BANDED SAND SNAKE. *Chilomeniscus cinctus.* 9–10 inches. Small, stout, blunt-nosed, smooth-bodied, sand-burrowing snake with numerous "saddles" or crossbands of black on a white, yellow, or reddish background. Below whitish. Nocturnal. Burrows in sand of dunes or in sand hummocks at the base of shrubs. Sonoran Desert.

47 ARIZONA LYRE SNAKE. *Trimorphodon lambda.* 2–3 feet. Rock-dwelling; ground-color, light brown, or gray above with brown dorsal patches, each with narrow lighter brown crossbars. V-shaped mark on head forming the "lyre." Western Arizona, south into Sonora.

48 WESTERN SHOVEL-NOSED SNAKE. *Chionactis occipitalis.* 10–16 inches. A small, ordinarily docile snake of sand or barren Lower Sonoran desert, with flat-topped snout and usually with 21 or more dark brown or black bands adorning the smooth-scaled white or yellowish body. The lower jaw is distinctly "countersunk" within the upper jaw. Mohave and Colorado deserts of California southward a short distance into Baja California and extreme northwestern Sonora. The SONORA SHOVEL-NOSED SNAKE (*Chionactis palarostris*), with convex snout and body bands usually fewer than 21, is found in western Sonora southward to the central portion of that state.

49 LEAF-NOSED SNAKE. *Phyllorhynchus decurtatus.* 12–20 inches. A small, snake; light-tan, pinkish, or gray-brown above, with 17 or more irregular, brown back-blotches; white to cream below. There is a distinct large scale coming up over the end of the "nose." Eyes large with vertical elliptical pupil. Active at night. May hiss when annoyed. Southern Nevada through southeastern California, southwestern Arizona, and desert portions of Sonora. In southern Arizona and desert parts of Sonora lives the SADDLED LEAF-NOSED SNAKE, with alternate narrow saddles of cream and broad saddles of brown or gray-brown.

50 GLOSSY SNAKE. *Arizona elegans.* 2½–4½ feet. A gentle, smooth-scaled snake of the creosote bush and mesquite deserts from western Texas south to Coahuila and west to southern California. The lower jaw is countersunk in the upper as in the shovel-nosed snake. The general ground color varies from yellowish gray to brown, buffy, or cream. Black-edged gray or reddish-brown blotches adorn the back. Sometimes confused with the common GOPHER SNAKE (*Pituophis catinefer*), a snake darker in color and possessing keeled scales.

51 CALIFORNIA BOA. *Lichanura roseofusca.* 2–3 feet. A heavy, smooth-bodied snake of slaty, bluish, or brownish-gray color with three broad lengthwise stripes of reddish brown. Yellowish white below with spots or blotches of brown. Docile, slow in movement. Nocturnal and crepuscular. Foothill and rocky areas of Mohave and Colorado deserts below 4,500 feet.

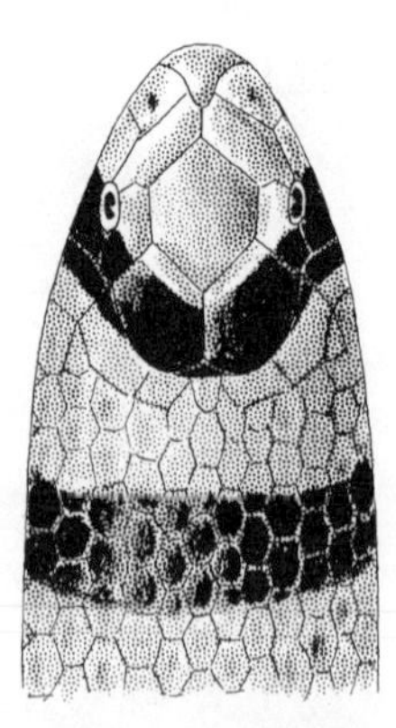

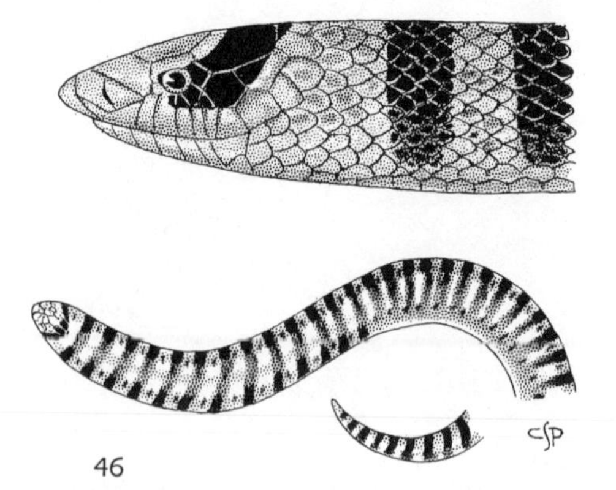

46

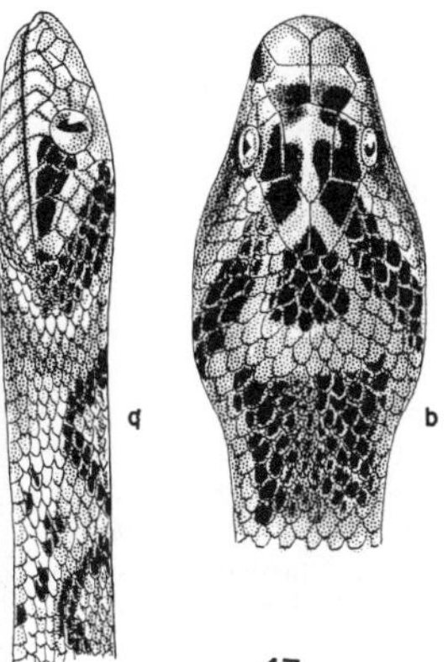

47

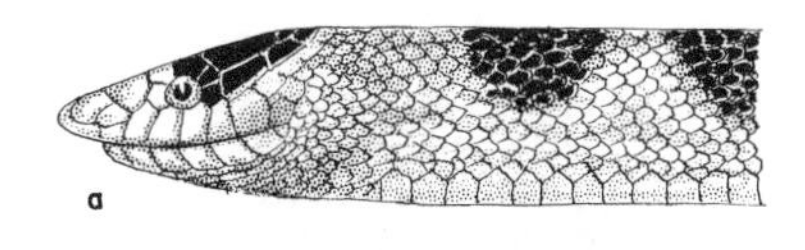

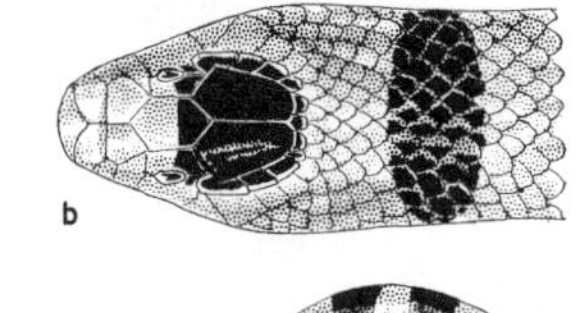

48

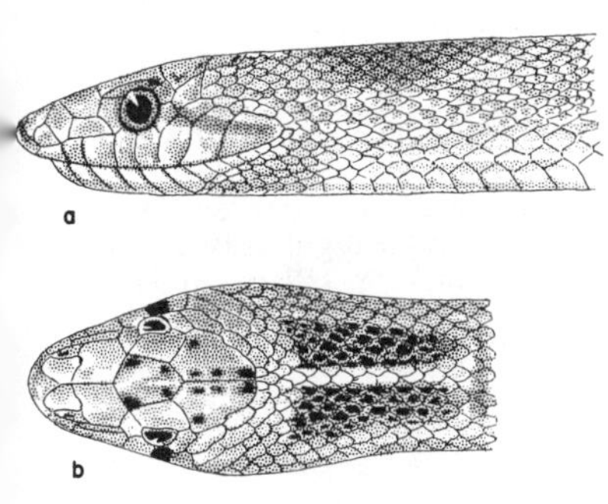

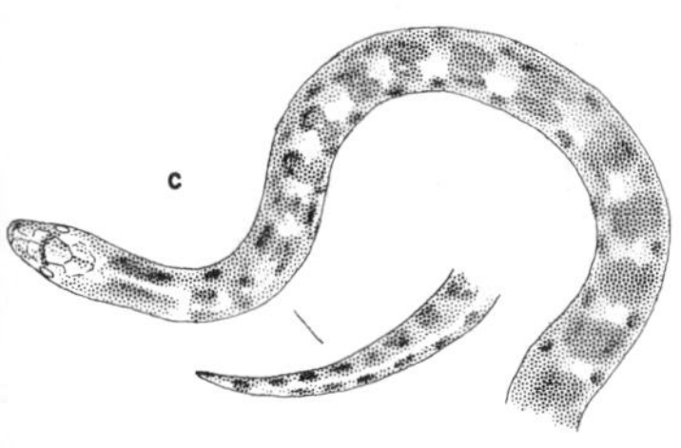

49

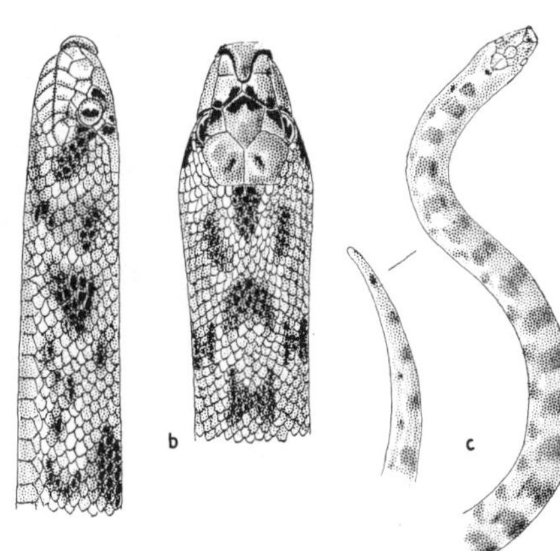

50

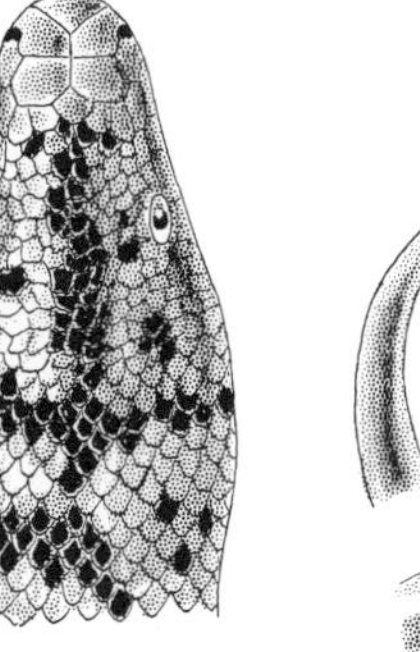

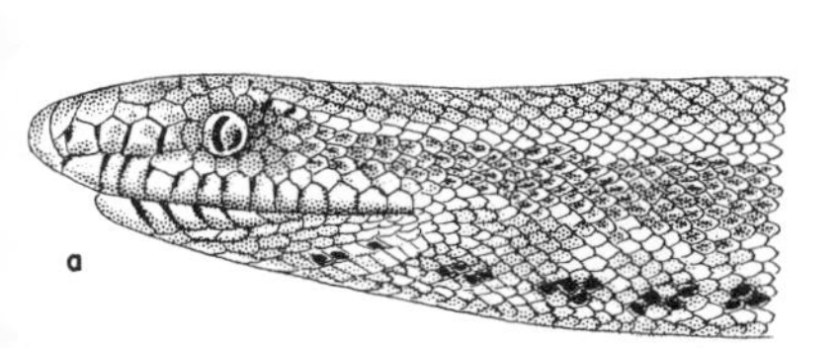

c

CSP

51

52 SPOTTED NIGHT SNAKE. *Hypsiglena torquata.* 12–18 inches. Slender with somewhat spade-shaped head. The ground color is gray or yellowish with numerous (52 to 54) dark-brown spots along back and sides; two large blotches on neck, these sometimes united above. Nocturnal and crepuscular. Feeds on lizards and insects. Lives among or near rocks of low desert and desert foothills.

53 DIAMOND-BACK RATTLESNAKES. *Crotalus sp.* 2½–5½ feet. Heavy-bodied venomous snakes with head distinct from slender neck; jointed horny rattle at end of tail; heat-sensitive pit between eye and end of nose. Rattlesnakes are found in almost all kinds of desert environments, from sandy areas to rocky foothills and mountains. (The number of species is indeed many. The reader must consult special books such as Robert C. Stebbins, *Amphibians and Reptiles of Western North America,* if he wishes to make definite identification of the kinds he finds.) Illustrated here is the side-winder.

54 CRESTED LIZARD or DESERT IGUANA (*Dipsosaurus dorsalis.* 12–16 inches. Generally whitish color, overlaid by bars and red-brown lines; top and sides of tail marked with series of transversely arranged spots or cross-bars. A row of short pointed scales down the middle of the back form a barely noticeable "crest." When running, often raises itself up on hind legs, with tail used as balancer, appearing then like a miniature dinosaur. Diurnal. Food: vegetable material, especially creosote bush blossoms, and some insects. Range: usually sandy areas of Mohave and Sonoran deserts.

55 YUCCA NIGHT LIZARD. *Xantusia vigilis.* 3–4 inches. Dark brown or light tan, finely peppered with black dots; large plates on head; tail fat and abruptly tapering, acting as a food reservoir; general color darker during day, becoming lighter at night. Nocturnal. Food: small night-crawling insects, especially termites. Range: usually under bark or fallen trunks of Joshua trees on Mohave Desert; also in other yuccas south to northern Baja California.

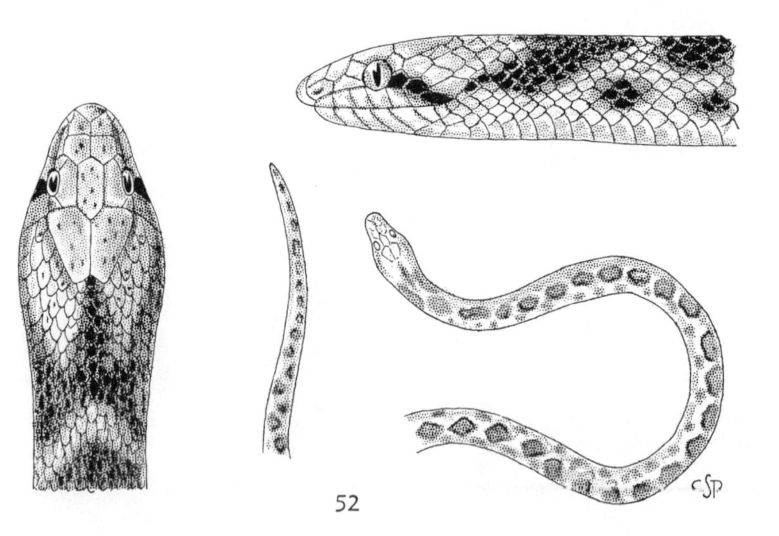

52

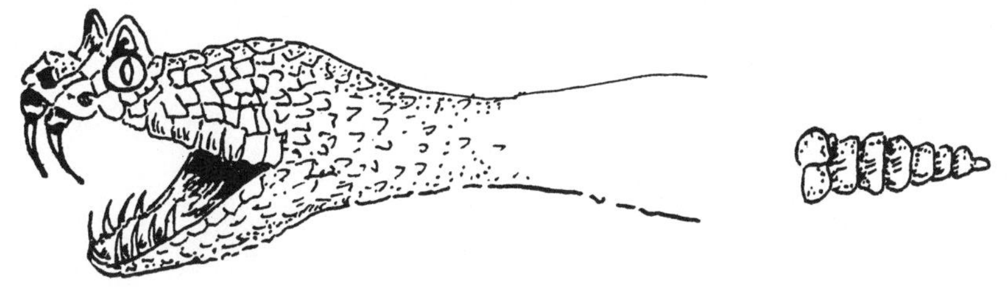

53

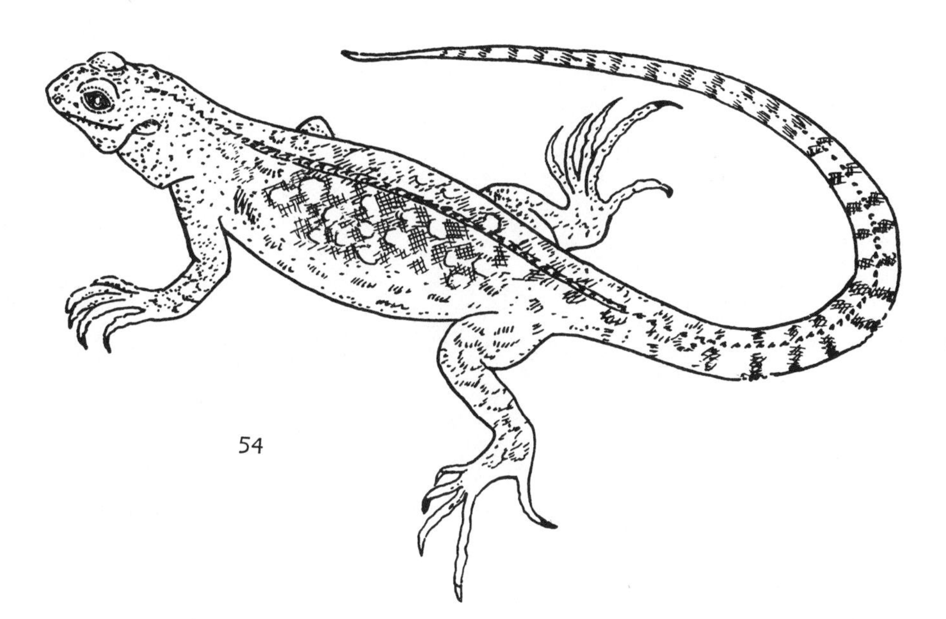

54

55

DESERT *Birds*

The desert's resident birds are largely insect- and spider-eaters, birds of prey, or scavengers. Theirs is a highly advantageous diet since it makes them to a large extent independent of ordinary sources of water such as are found at seeps, springs, and small canyon streamlets. Some of them may, or do, drink water when it is available but they are not seriously discommoded when deprived of it. Quail and such birds as the phainopepla are not necessarily dependent on water holes and streams either, since they are big berry eaters, and in the fruits of the leafless mistletoe and several kinds of cacti and the succulent red tomatillos of the shrubby lyciums they find sufficient fluid to maintain a proper body water balance. The several species of hummingbirds are nectar feeders and are often seen during the season when bright-blossomed shrubs are in evidence.

The larger birds of prey such as the golden eagle, numerous hawks, falcons, and owls are often uncommonly successful desert dwellers because they can get water sufficient for most of their needs from the body fluids of the reptiles, birds, and mammals which they kill. The scavengers such as the caracaras and ravens eat, besides carrion, living lizards, snakes, and insects.

Desert birds are usually easy to identify because they are so easily and advantageously seen. Unlike forest-dwelling birds, often discouragingly hidden by dense foliage, both when still and in flight, the desert birds are usually seen out in the open or sitting in thin-foliaged shrubs or trees or on top of cacti or barren tree limbs. Their nests too are on the whole more exposed and easily located.

In the spring and again in autumn many migrant birds on their way north or south stop to rest and drink at the desert streams and scattered water holes. There they are more advantageously seen than when in their usual environment in woods and mountain brush-covered areas.

56 PACIFIC HORNED OWL. *Bubo virginianus pacificus.* Length 18–24 inches. This is the desert's "hoot owl," its deep booming notes most often heard in canyons, rocky hills and tree-lined washes. A fearless hunter of small mammals. The throat is white; the rest of the underparts yellowish white, finely barred with black; upper parts mottled with black, white, and light brown. California and Arizona. A paler subspecies (*pallescens*) inhabits arid parts of the Vizcaíno Desert of Baja California.

57 WESTERN BURROWING OWL. *Speotyto cunicularia hypogaea.* Length 9 inches. A small brown, long-legged, earless owl often seen in the daytime near its hole in earth banks or on level deserts. Often badger holes are occupied. It is a cricket and grasshopper feeder for the most part, but also may eat squirrels, lizards, mice, scorpions, and centipedes. Widely scattered over our deserts.

58 BARN OWL. *Strix pratincola.* Length 14–15 inches. A large hornless owl with prominent facial disks. The upper parts are buffy, speckled finely with black and white; the lower parts white to buff minutely speckled with brown. Widely distributed over many parts of the world, including our deserts.

59 LUCY'S WARBLER. *Vermivora luciae.* Length 4 inches. A small, very plain, gray and buffy bird with white underparts and distinct patch of rich chestnut on the rump. In summer there is a definite unconcealed crown patch of chestnut. Breeds in the far western deserts of the United States and northern Baja California. Sometimes called the DESERT WARBLER because of its close association with arid lands.

60 ELF OWL. *Micrathene whitneyi.* Length 5–6 inches. A sparrow-size, earless bird, with close association with giant cacti. Above grayish brown; below whitish streaked with vertical blotches of dark-brown or reddish-brown. The eyebrow feathers are white. From south Texas to southern California, Baja California and the Mexican tablelands.

61 LONG-EARED OWL. *Asio wilsonianus.* Length 14–16 inches. A slender, graceful, long-winged owl with conspicuous ear tufts. A night hunter remaining during the day concealed in part by the foliage of trees in which it roosts. Above deep brown mottled with tawny; abdomen whitish streaked and barred with brown. Note: a soft, deep-toned *hoo-oo,* snarling *eee-aaaa-o-o-oow,* or thin nasal *eeee-uuuuh*!

62 ARIZONA CRESTED FLYCATCHER. *Myiarchus tyrannulus.* Length 9–9½ inches. A tyrannid much resembling the ASH-THROATED FLYCATCHER, but a little longer. The beak is heavier and the call note harsher and given with more vigor. Both birds have yellow on the breast. Deserts of southern Arizona, Rio Grande Valley in Texas, and southward through Mexico, except Baja California.

63 SAY PHOEBE. *Sayornis saya.* Length 7¾ inches. A familiar bird of our southwestern deserts in spring and summer; in winter it is seen in the central Mexican plateau. Upper parts grayish; breast and abdomen bufly; tail dull black. The plaintive song once heard is readily recognized again. Nests are often built in old mine shafts, wells, and miners' cabins.

64 ASH-THROATED FLYCATCHER. *Myiarchus cinerascens.* Length 7½–8 inches. A common desert resident from the Great Basin south into Mexico's central plateau and Baja California. A proud, handsome bird with prominent black eyes. Upper parts grayish-brown; throat and breast pale ash; belly pale sulphur yellow. Nests in old woodpecker holes of giant cacti, in knotholes of mesquites, and in stump cavities.

18 - 24"
9"
14 - 15"
56
57
58
14 - 16"
5 - 6"
60
61
59
4"
7 3/4"
63
62
9 - 9 1/2"
7 1/2 - 8"
64

65 AUDUBON CARACARA. *Polyborus cheriway.* Length 20–24 inches. A singular-looking, falcon-like bird with black-crested head; throat and neck white, the latter barred with black. Bare portion of face bright orange. Abdomen blackish; pale patches on wing tips. Tail with terminal black band. A bold hunter both on the wing and on foot. Often observed alone or in small groups atop organ pipe and sahuaro cacti and ironwood trees of the Sonoran plains. Found also in Baja California.

66 PRAIRIE FALCON. *Falco mexicanus.* Length 17 inches. A speedy, bold-flying, sand-colored hawk with pointed wings. The throat is white, the other underparts buffy white streaked with sooty brown. Arid parts of western United States and Mexico. The prairie falcon builds its nest high on rock ledges or in holes of vertical cliffs. The cry is a high-pitched, sharp *kee, kee, kee.*

67 DESERT SPARROW HAWK. *Falco sparverius deserticola.* Length 10½ inches. A small, bright-colored hawk often seen in flight or perching on fence wires or posts. The back is rich reddish-brown barred slightly with black; the head is slate blue with reddish-brown crown; wings slate blue. The primaries barred with white. Prominent are the black spots of light buff to reddish-brown on breast and abdomen. It has a high-pitched note: *killy, killy, killy.*

68 RED-TAILED HAWK. *Buteo borealis.* Length 18–23 inches. A large hawk sometimes taken for an eagle by the novice in bird knowledge. The best field mark is the rufous-red of the upper tail. The wings are broad, the tail broad and rounded. Wide-ranging, and one of the largest commonly seen desert hawks. Erroneously persecuted as a "chicken hawk."

65

17"
66
10 1/2"
67
18 - 23"
68

69 TURKEY VULTURE. *Cathartes aura.* Length 28–30 inches. A sooty-brown or blackish scavenger known to the Mexicans as the zapolote and to Americans as the turkey buzzard. The bare skin of the neck and head are dull purplish-red, the beak whitish. The BLACK VULTURE (*Coragyps atratus*), widely spread in Mexico, is somewhat smaller and with blackish beak, head and neck. The black vulture flaps its wings frequently while the turkey vulture is more given to soaring.

70 SCALED QUAIL. *Callipepla squamata.* Length 10–12 inches. Sometimes called "blue quail" because of its general bluish-gray color and "cottontop quail" because of its unique white-tipped, tufted crest. The name "scaled quail" refers to the distinctive scale-like marking of the mantle, breast, and sides. Arid lands of central and southern Arizona to west central and southern Texas; south through Sonora and the Chihuahuan Desert.

71 HARRIS HAWK. *Parabuteo unicinctus harrisi.* Length 22 inches. A large chocolate-brown to black hawk, with white on rump and white band at tip of tail. Southeastern California, southern Arizona, southern New Mexico, southern Texas; south into Mexico and Baja California.

72 MEARNS' QUAIL. *Cyrtonyx montezumae mearnsi.* Length about 8 inches. A small quail found mostly in the shrub-dotted slopes of arid mountains of central Arizona and New Mexico and west Texas; northwestern Mexico exclusive of Baja California. The facial markings are unique.

73 GAMBEL'S QUAIL. *Lophortyx gambeli.* Length 10–12 inches. A desert-dwelling quail with overhanging plume of black; black forehead and throat. The chestnut flanks and sides are prominently striped with white. Mohave and Colorado deserts east to extreme west Texas; northeastern Baja California, Sonora, Sinaloa, Tiburón Island.

70

71

72

73

74 PINON JAY. *Cyanocephalus cyanocephalus.* Length 10–11½ inches. A rather plain-looking, non-crested, grayish-blue jay, with long, sharp beak and rather short tail. The throat is streaked with white. In many ways it acts like a crow. Especially in autumn it is seen in the piñon-juniper desert uplands, consorting in comparatively large flocks. Its notes are not harsh like those of most jays. Arid Upper Sonoran Zone of western United States and northern Mexico where piñons and junipers abound.

75 MAGPIE JAY. *Calocitta formosa.* Length 20–25 inches. A particularly handsome artistocrat among jays. The prominent crest and long tail, together with the black band of the neck are distinctive characters. Bold and inquisitive and with unmusical whistling note. Arid lowlands of Mexico along Pacific Coast; also south central Mexico.

76 GILA WOODPECKER. *Centurus uropygialis.* Length 8–9½ inches. A bird of the tree-cactus country, with marked barrings of black and white on the back, wings, rump, and tail. The head is gray with crown-patch of red, the lower abdomen yellow. The white rump-patch is conspicuous in flight. Sometimes called the gray-breasted woodpecker. Feeds on mistletoe berries, cactus fruits, and insects. Dry parts of southern Arizona, southwestern New Mexico, along the Colorado River in southeastern California, and southward into Baja California, Sonora, and the southwestern Chihuahuan Desert.

77 GILDED FLICKER. *Colaptes chrysoides.* Length 10–11 inches. Much like the red-shafted flicker but the feather shafts and underside of wings and tail are bright yellow. Often found among the giant and other tree cacti of southern Arizona, Sonora, and northern Baja California, south to latitude 30°. Note the red "mustache." Nests are found in hollows made in giant cacti and cottonwood trees.

78 LADDER-BACKED WOODPECKER. *Dryobates scalaris.* Length 7–7¾ inches. The common name refers to the black and white bars of the upper parts. The outer tail feathers are also barred but the middle tail feathers are wholly black. A resident of the lower deserts both in the United States and Mexico. It nests in holes made in yuccas, agaves, mesquites, and other desert trees. Known also as the Texas woodpecker.

79 ROADRUNNER. *Geococcyx californianus.* Length about 22 inches. Sometimes called "chaparral cock" because of its preference for brush-covered areas, and "ground cuckoo" because of its relation to the common cuckoos and its ground-loving habits. It seldom flies, preferring, unless hard pressed, to run. The crest can be raised or lowered at will. Southwestern Kansas and southern Colorado to central California and Nevada; southward to central Mexico and Baja California. In the arid lowlands of western and central Mexico is found the smaller LESSER ROADRUNNER (17–20 inches), with buff-colored underparts; underside of neck, and middle breast, without stripes.

80 WHITE-THROATED SWIFT. *Aeronautes saxatilis.* Length 6½–7 inches. A bird generally seen while feeding on the wing. Striking are the long, narrow wings and sharply contrasting black and white underparts. The tail is comparatively long and slightly notched, but without spines at the ends of the terminal shafts. The flight is bold and exceedingly swift. The nests are placed in crevices of cliffs. Western United States, south to Mexico.

81 COOPER'S TANAGER. *Piranga rubra.* Length 7–7½ inches. A magnificent bird. The male is almost wholly rose-red, brightest on the crown and on the underparts. The female is less spectacular, with yellow and yellowish-olive predominating. A bird that is usually seen along river bottoms of arid southeastern California and southern Nevada, eastward to west Texas.

10-11 1/2"
74
20-25"
75
81
7-7 1/2"
8-9 1/2"
10-11"
76
77
7-7 3/4"
78
22"
79
6 1/2-7"
80

82 PHAINOPEPLA. *Phainopepla nitens.* Length 7–7½ inches. Slender, glossy-black bird, with head having a conspicuous crest; white wing patches much in evidence in flight. The phainopepla is most often seen atop trees such as mesquite, desert willow, and ironwood which harbor the leafless mistletoe upon whose berries the bird heavily feeds. In arid lowlands of southern Nevada, southern California, southern Arizona, Sonora, Chihuahua, and Baja California.

83 NUTTALL'S POORWILL. *Phalaenoptilus nuttalli.* Length 7–8 inches. Smaller and lighter colored, but somewhat similar to the EASTERN WHIPPOORWILL. The oft-repeated call note lacks the "whip" of the eastern bird. Prominent are the large white throat patch and outer tail feathers tipped with white. Recent observations show that this bird hibernates in winter. It is often seen flying low or sitting in the dust at nightfall. Arid areas from Canada to Mexico.

84 WHITE-NECKED RAVEN. *Corvus cryptoleucus.* Length 18–21 inches. When sitting still it looks like a very large crow but it is different in flight. The raven alternates flapping its wings and soaring; the crow flies with rapid wing-beat and never soars. The voice of the raven is a definite *croak*; that of the crow a *caw*. Ravens are generally seen alone, in pairs, or in small family groups except in early spring, when large numbers (20 to 40) gather for play-flight exhibitions. Crows are usually seen in flocks and seldom on deserts. The white-necked raven is seen in the Lower Sonoran Zone of Texas west to California and south into northern Mexico. The name refers to the white at the base of the neck feathers.

85 WHITE-RUMPED SHRIKE. *Lanius ludovicianus.* Length 9 inches. A big-headed, gray-white-and-black bird with narrow tail and large head. A broad bar of black over sides of head; black on wings and upper tail. Generally seen sitting alone atop sentinel post on yuccas, ocotillos, or other shrubs of considerable height. Breeds over much of western United States and northern Mexico. Winters in arid Southwest and Mexico.

82

7-8"
83
18-21"
84
9"
85

86 HORNED LARK. *Otocoris alpestris.* Length 7–8 inches. A ground-loving, walking bird, but at times it rises in easy graceful flight and circles about before again lighting. The size is somewhat larger than that of a sparrow. The head is adorned with two small black erectile horns joined by a black band. A black band extends also across the chest just above the white breast and abdomen. The throat is bright yellow, the tail black. In the autumn and winter horned larks are seen in flocks on the desert floor where they seek seeds. They nest on the ground. Widespread on all deserts.

87 TEXAS NIGHTHAWK. *Chordeiles acutipennis texensis.* Length 7½–8½ inches. A dull mottled-gray bird, with white throat and distinct white wing bars. While feeding, flies rather close to the ground. Because of its note it is sometimes called the trilling nighthawk. It does not "boom." Widely distributed in the deserts of the United States and Mexico, including Baja California.

88 PLUMBEUS GNATCATCHER. *Polioptila melanura.* Length 4½ inches. A small leaden-gray bird, with black cap (in male) and dark tail, having the web of the outer feathers white. A lively, fidgety bird frequenting bushes of the Colorado and, less often, Mohave deserts. The song is a somewhat chickadee-like *tsee-dee-dee-dee-dee*, the call note a *chee-chee-chee.* Nest, a compact cup-like structure of plant fibers built in trees and shrubs.

89 LECONTE'S THRASHER. *Toxostoma lecontei lecontei.* Length 10–10¾ inches. A pale grayish-brown bird, with dark tail. Long, slender, much curved beak. A fine singer, often giving a whistling note. Generally seen skulking along the ground or rapidly running from bush to bush. With two other thrashers it shares the southwestern deserts of the United States and northwestern deserts of Mexico. Bendire's thrasher has a shorter, less bent beak. The Crissal thrasher has a decurved beak like Leconte's thrasher but has an over-all darker color and chestnut undertail coverts.

90 SAGE THRASHER. *Oreoscoptes montanus.* Length 8 inches. The grayish-brown, short-tailed, straight-beaked thrasher of the sagebrush country.

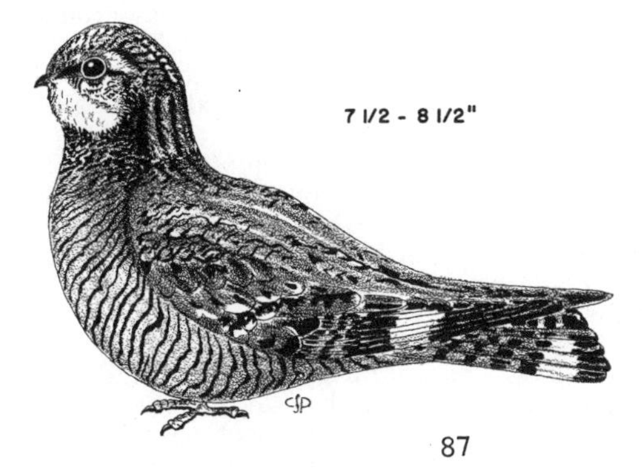

90

91 WHITE-WINGED DOVE. *Melopelia leucoptera.* Length 10–11 inches. This beautiful grayish-brown dove is somewhat heavier than but near the size of the MOURNING DOVE. Unlike that bird it has a rounded tail tipped with white. The adult male has a large patch of white on the wing coverts. A widely spread species both in Mexico and in the deserts of southeastern California, southern Arizona, southwest New Mexico, and southern Texas.

92 MEXICAN GROUND DOVE. *Columbigallina passerina.* Length 6–6½ inches. A very small, almost sparrow-size dove, with short black tail and wings showing a flash of rufous when in flight. The lower throat and breast are more or less spotted with black. Arid regions of Mexico generally, but less frequent on the central plateau; also southeastern California, southern Arizona and east to western Texas.

93 MOURNING DOVE. *Zenaidura macroura.* Length 11–13 inches. Widespread over the deserts of the southwestern United States and Mexico. This long-tailed, graceful bird is most frequent near sources of water such as springs and streams. However, it may feed and nest at a considerable distance from water. The pointed tail is a distinguishing field mark.

94 COSTA'S HUMMINGBIRD. *Calypte costae.* Length 3½ inches. Both forehead and throat of the male are amethyst or purple, changing to green or blue, and the feathers of the ruff or gorget stand out noticeably. The back is greenish-bronze. Resident in Baja California and Sonora. Winters south to Sinaloa. Costa's "hummer" is known for the high-pitched, hissing sound made by the male as it dives through the air at the time of courting. Seen during the breeding season in deserts of southern California, southern Utah, Nevada, and Arizona.

95 INCA DOVE. *Scardafella inca inca.* Length 7½–8 inches. A very tame bird of the open desert; also known as the "scaled dove" because of its scaled plumage. The tail is long and white-sided. General color gray to grayish-brown. The primary wing feathers are chestnut tipped and edged with sooty black. The legs are pink. Inca doves have charming manners and are much given to making many accented cooing notes as if talking among themselves.

96 BLACK-CHINNED HUMMINGBIRD. *Archilochus alexandri.* Length 3¾ inches. Male readily recognized by its velvety black throat and white collar just beneath; above metallic bronze-green, tail deep purplish. A bright purplish-violet patch is seen in certain lights on the throat. The male performs an aerial courting dance. Deserts of southeastern California; also often seen in southern Arizona, New Mexico, and west Texas, northern Baja California and Sonora.

97 ALLEN'S HUMMINGBIRD. *Selasphorus alleni.* Length 3⅓ inches. The male has a red throat, green back, and rufous rump. Seen occasionally in early spring in southeastern California deserts. It winters in Baja California and the northwestern Mexican mainland.

10-11"
91
6 - 6 1/2"
92
11 - 13"
93
3 1/2"
94
3 3/4"
96
7 1/2 - 8"
95
3 1/3"
97

98 PYRRHULOXIA. *Pyrrhuloxia sinuata.* A beautiful crested, aristocratic near-relative of the ARIZONA CARDINAL. The upper parts are light grayish brown, but a ring around the beak, the throat, breast patch, thighs, and undersides of wings are bright rosy red or vermilion. The crest is tipped with darker red. The beak is parrot-like and in summer is yellow or orange. The nest is built in thorny trees or shrubs. Lower arid areas from Arizona to west Texas, and northern and western Mexico and southern Baja California.

99 CACTUS WREN. *Heleodytes brunneicapillus couesi.* Length 7–8 inches. Largest of our desert wrens, being from one to two inches longer than any in its area. The heavily spotted throat and breast and chuttering note are distinguishing characters. The large purse-like nest of grass and twigs of desert annuals is generally built amid protecting spines of chollas. Note the white line over the eye. Mohave and Colorado deserts east to southern Texas and south over the Mexican plateau to Jalisco; also on the Yucatán Peninsula and in Sonora and Baja California.

100 WESTERN MOCKINGBIRD. *Mimus polyglottus leucopterus.* Length 10–12 inches. A slender, robin-sized, grayish and clay colored bird of varied vivacious and imitative songs; often seen singing atop a yucca, agave stalk, or desert shrub. As it flies upward conspicuous white wing-patches are seen. Often seen on deserts of southwestern United States, the Mexican plateau, and in Baja California.

101 VERDIN. *Auriparus flaviceps.* Length 3½–4 inches. A small, very active bird with ash-gray or ashy-brown body; the head bright yellow or olive-yellow. The bulky nest, built of small twigs and lined with feathers, is generally placed rather low in thorny trees or shrubs. It is entered by a small round hole at one end. The nests may be used in winter for sleeping quarters. Northern and western Mexico, northern Baja California; Colorado and Mohave deserts in California; southern Arizona east to Texas along the Rio Grande.

102 ROCK WREN. *Salpinctes obsoletus obsoletus.* Length 5–5¾ inches. A pale grayish-brown wren with minute streaks and speckles of black, white, and brown. The tail is tipped with buffy brown and has an inner band of black. Distinctly an inhabitant of arid rocky hills and mountain sides. It builds its nest in crypts in rocks and always has from few to many small flat stones near the entrance. The tinkling, spirited song is most appealing. Its food consists of small insects and spiders. Arid portions of western United States and similar habitats in the Mexican deserts.

98
8 3/4"
7 - 8"
99
10 - 12"
100
5 - 5 3/4"
101
3 1/2 - 4"
102

103 SAGE SPARROW. *Amphispiza nevadensis.* Length 5½–6¾ inches. A gray bird with prominent dark breast-spot and broken black "whisker marks" below white band on throat. The sage sparrow is similar to BELL'S SPARROW and often occurs with it. The latter has black marks on side of throat broader and unbroken. Both birds winter in the southern United States deserts and northern parts of Baja California, Sonora, and Chihuahua.

104 DESERT SPARROW. *Amphispiza bilineata.* Length 4¾–5 inches. No sparrow breeding in desert territory has a more cheery sweet song than this gray-backed bird with distinctive white face-stripes and black throat-patch. In Mexico, found in Baja California and desert areas of the mainland south to Sinaloa, Durango, and Hidalgo. Widespread in desert areas of the United States.

105 WESTERN LARK SPARROW. *Chondestes grammacus strigatus.* Length 6–6¼ inches. A bird of easy identification because of its boldly patterned head and white-tipped, rounded tail. Often seen in winter on the deserts of California, Arizona, southern Texas, northern Chihuahua and Baja California. A musical, fervent, and continuous singer.

106 ABERT'S TOWHEE. *Pipilo aberti.* Length 8½–9 inches. A shy, pale grayish-brown towhee of rather large size. A black to dusky patch around the base of the beak is a distinguishing mark. Frequents mesquite and cottonwood thickets of southeastern California deserts to Arizona and New Mexico, northeastern Baja California, and northwestern Sonora.

107 GAMBEL'S WHITE-CROWNED SPARROW. *Zonotrichia leucophrys gambeli.* Length 6–6¼ inches. A winter resident of southern United States deserts, through Baja California and Sonora and Mexican Plateau. The black-and-white-striped crown and pinkish bill are distinctive. The song, a series of clear plaintive notes, is followed by a "trilled whistle." This handsome sparrow breeds in the high mountains of western United States and Canada.

108 BLACK-CHINNED SPARROW. *Spizella atrogularis.* Length 5¼ inches. A pinkish-billed, gray-headed bird with black on upper throat and ring around beak. The back is rusty brown or cinnamon, streaked with black. Nests in the deserts of California, Arizona, and southern New Mexico, and southward to northern Baja California, northeastern Sonora, and the eastern and southern parts of the Chihuahuan Desert.

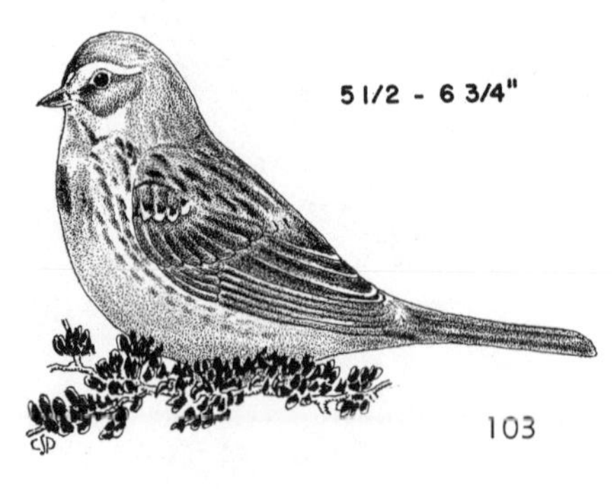

105

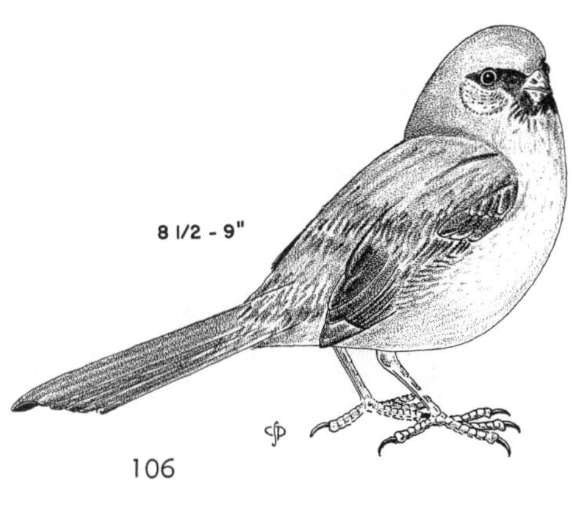

106

107

108

109 ARKANSAS GOLDFINCH. *Spinus psaltria.* Length 4 inches. A small finch with yellow breast, black cheeks and forehead, black-and-white wings. The back is olive-colored to black, depending on age. This bird was first described from specimens taken on the Arkansas River in Colorado, hence the name. The GREEN-BACKED GOLDFINCH, commonly so called because even the immature males have olive-green backs, is considered to be a different subspecies. Often seen on deserts of southeastern California and western Arizona and southward into Mexico.

110 VERMILION FLYCATCHER. *Pyrocephalus rubinus.* Length 5–5½ inches. The brilliantly colored male is always a startling sight and is one of the most characteristic birds of the dry Mexican desert and of arid areas of southern New Mexico, southern Arizona, and western Texas. The upper parts of the male are dusky brown but the top of the head and underparts are bright scarlet. A frequenter of mesquite bosques, streamside cottonwoods, and willows.

111 WESTERN MEADOW LARK. *Sturnella neglecta.* Length 8–10 inches. A rather large, chunky brown bird, with short wide tail with showy white patch on either side and bars on the middle feathers. Under parts yellowish with black crescent mark on throat. A beautiful bird with clear, sweet, bubbly song. Distributed from western Canada south to Baja California and northern Sonora and Chihuahua. Winters south to Jalisco.

112 SCOTT'S ORIOLE. *Icterus parisorum.* Length 7½–8 inches. From west Texas to the deserts of southern California and southward into Mexico, this handsome oriole can be found in yucca, agave, and sotol country. The song is of clear meadow-lark quality. The body is mainly black except for the white wing-bars and lemon yellow of belly, shoulders, rump, and basal half of lateral tail feathers.

113 BULLOCK'S ORIOLE. *Icterus bullocki.* Length 7–8 inches. A yellow-faced oriole with black crown, back, and upper throat; under parts (except throat) bright orange. Breeds from the Mohave Desert eastward through southern Arizona and southwestern New Mexico, southward to northern Baja California; also northwestern Mexico, south to Tepic. Nests, cup-shaped and semipensile, often seen in palms, figs, and sycamores.

111

112

113

DESERT *Mammals*

In speaking of mammals it must be said that the rodents (ground squirrels, wild mice, kangaroo rats, etc.) outnumber all other kinds. Many animals, such as jack rabbits and certain wild mice are not at all dependent on free water of rainwater pools, springs, or streams, since they are able to synthesize the water they need from the chemical elements found in their food.

Says Dr. Knut Schmidt Nielson, referring to small rodents: "Even the driest seeds contain some absorbed water, but a larger quantity is formed by oxidation of food in the body. On oxidation one gram of starch yields 0.6 gram of water and one gram of fat yields almost 1.1 grams of water. By exercising the greatest physiological economy with water for urine and for feces and for evaporation (which cannot be completely avoided because the expired air is saturated with water vapor) these small rodents can just manage on the oxidation water, being independent of intake of free water.

"These small mammals do not use water for heat regulation. They are nocturnal and remain in their underground burrows throughout the hottest part of the day. They are an ecological paradox, living in the desert without being exposed to the rigor of desert heat.

"The large animals cannot escape the desert heat by hiding underground. To avoid undue rise in the body temperature they evaporate water from the surface of the skin (sweating) or from moist respiratory surfaces (panting)."*

Coyotes, foxes, bobcats, and skunks are the principal carnivores. They drink at the water holes when possible, but depend

* "Animals and Arid Conditions," in *The Future of Arid Lands* (Washington, D.C.: American Association for the Advancement of Science, 1956), pp. 371–72.

much for their liquid intake on the blood of insects, birds, reptiles, and mammals, which are their chief food. They are particularly successful beasts either because of their cunning, swiftness of foot, or night hunting habits.

The largest desert mammal is the bighorn sheep. Once plentiful, it has fallen on evil days as the number of human hunters have increased and its wild habitat has been invaded by human dwellers. Bighorn have been able to hold their own or slightly increase their numbers only in sanctuaries or refuges set aside for them in lonely steep-sided mountains. This noble mammal is especially adapted for moving over country of rough terrain and it has unusual ability to seek out and utilize nearly inaccessible and quite isolated sources of water.

By far the smallest and probably least known of all desert mammals is the desert shrew (*Sorex crawfordi*). Its ashy-gray body is scarcely more than two inches long, to which is added about an inch-long, smooth mouse-like tail. It is a fierce little beast and during its waking hours does little else than hunt for food (insects, larvae, and small rodents) among the grasses and leaves around and under shrubs of its arid habitat.

All desert mammals have fur coats of lighter colors than those worn by their representatives living in more humid environments. The brown colors of the mountain- and coastal-dwelling mammals are replaced by fawns and tans, and dark grays by cinereous and lighter grays.

114 BLACKTAIL JACK RABBIT or DESERT HARE. *Lepus californicus.* 17–21 inches, head and body; 6–7 inch ear. Gray-brown, with large black-tipped ears and black streak on top of tail. Mostly nocturnal. Food: green vegetation, bark. Often seen running across roads at night, when many are killed by cars. Range: all five of our deserts. The white-tailed, longer-eared ANTELOPE JACK RABBIT (*Lepus alleni*) does not have the black tips on the ears, the head is smaller, and the tail is not topped with black; it is found only from south central Arizona to southern Sonora, and in northwestern Chihuahua.

115 DESERT COTTONTAIL. *Sylvilagus auduboni.* 12–15 inches, head and body; 3–4 inch ear. Pale gray, with prominent white tail. Digs burrows. Food: fresh vegetation and bark. Nocturnal but may be seen sometimes during the day. Often noticed hopping across road at night or early morning; many are hit by cars. Range: all five of our deserts, except western Great Basin Desert.

116 ROCK SQUIRREL. *Citellus variegatus.* 10–11 inches, head and body; 7–10 inch tail. Largest of the desert's ground squirrels; dark mottled gray, with slightly bushy tail. Diurnal. Hibernates. Usually lives around or near rocks; most often seen on top of boulders. Food: fresh vegetation, seeds. Call note: sharp, loud *chirp.* Range: Chihuahuan Desert north and westward to the Navahoan and Great Basin deserts.

117 ROUND-TAILED GROUND SQUIRREL. *Citellus tereticaudus.* 5–6 inches, head and body; 2–4 inch tail. Pinkish cinnamon on a background of gray; tail pencil-like, not bushy; ears exceedingly small; no contrasting markings. Food: seeds, foliage. Diurnal. Hibernates. Burrows: about 2 inches across in hard-packed soil, usually in the open. Often seen running across road during heat of summer day. Range: Mohave and Colorado deserts, and Sonoran deserts of southern Arizona and western Sonora.

118 ANTELOPE GROUND SQUIRREL or DESERT "CHIPMUNK." *Ammospermophilus leucurus.* 5–6 inches, head and body; 2–3 inch tail. Prominent white line on each side of back, ending at shoulder; tail, which is usually carried curled over back, has a prominent white underside. Often glimpsed running across road. Our only ground squirrel that does not hibernate. Call: a series of sharp birdlike notes or a rapid chattering. Food: seeds, foliage, small insects. Diurnal. Burrows: small round openings about 2 inches across, generally among rocks or roots at base of a shrub. Range: gravelly open stretches between shrubs and rocks over most of the North American deserts except in southern Arizona, southwestern New Mexico, and Sonora, where its place is taken by the YUMA ANTELOPE GROUND SQUIRREL (*Citellus harrisi*), which is slightly larger and has a longer gray tail with no white underneath.

114

115

116

117

118

119 DESERT WOOD RAT, PACK RAT, TRADE RAT, or BRUSH RAT. *Neotoma lepida.* 5–7 inches, head and body; 4–6 inch tail. Gray to brownish color, grayish underneath; white feet; tail brown-gray above and white below. Nocturnal. Constructs nests or "houses" of twigs, cactus, and small rocks at base of bush, under a rock, or in an open sandy area; through its often large bulky "home" there are many passageways, the entrances protected with cactus joints when available. Food: seeds, foliage, small insects. Range: over most of our deserts except the Chihuahuan and in southeastern Arizona and most of Sonora. The WHITE-THROATED WOOD RAT (*Neotoma albigula*), with a distinctly white throat instead of gray, occurs over all the deserts except most of Baja California and the Great Basin Desert. The MEXICAN WOOD RAT (*Neotoma mexicana*), usually found only in rocky areas, is generally darker in color with a bicolor tail (black above, white below) and occurs only in the eastern half of our deserts: high deserts of southeastern Utah, eastern Arizona, western New Mexico, parts of southern Colorado, and northwestern Chihuahua.

120 DESERT KANGAROO RAT. *Dipodomys deserti.* 5–6 inches, head and body; 7–8 inch tail. Light yellowish tan with white under parts; distinct black streak across the hips; tail white-tipped. This largest of the kangaroo rats has long balancing tail and four toes on feet of hind legs adapted for jumping. Nocturnal. Food: seeds, some green vegetation. Cheek pouches are used for temporary storage while gathering and transporting food. Range: sandy plains or dune areas of the Mohave and Colorado deserts, and similar areas of Yuman Desert. Other desert-dwelling species include the ORD, MOHAVE, MERRIAM, GREAT BASIN, and BANNERTAIL KANGAROO rats.

121 WHITE-FOOTED or DEER MOUSE. *Peromyscus, sp.* 3–4 inches, head and body; 2–6 inch tail. White feet; large rounded ears; big bulging black eyes; white belly; soft brownish back; tail relatively long. Nocturnal. Common species include the PINON, CACTUS, WHITE-ANKLED, ROCK, and BRUSH mouse. Food: seeds, vegetation, small insects. Burrow: about one inch across under rocks or at base of plant. Range: all six of our deserts; some species invade human habitations.

122 SPINY POCKET MOUSE. *Perognathus spinatus.* 3–4 inches, head and body; 3–4 inch tail. Pale yellowish color, with long tail ending in brush; spine-like hairs on rump. Nocturnal. Burrows: about 1½ inches across at base of shrub or rock. Food: seeds, fresh vegetation. Cheek pouches for aid in foraging. Range: Colorado Desert through Baja California. Closely allied species include the HISPID, BAILEY, ROCK, NELSON, APACHE, and LITTLE POCKET mouse. The last-named, *Perognathus longimembris*, is almost wholly restricted to the peculiar "desert pavement" areas of the Mohave, Great Basin, and Sonoran deserts.

123 VALLEY POCKET GOPHER. *Thomomys bottae.* 5–7 inches, head and body; 2–3¾ inch tail. Burrowing, brownish rodent of variable size and color (almost white in southern Colorado Desert and smallest in some of the southern desert mountains). Range: southern Great Basin, Sonoran, and northern Chihuahuan Desert.

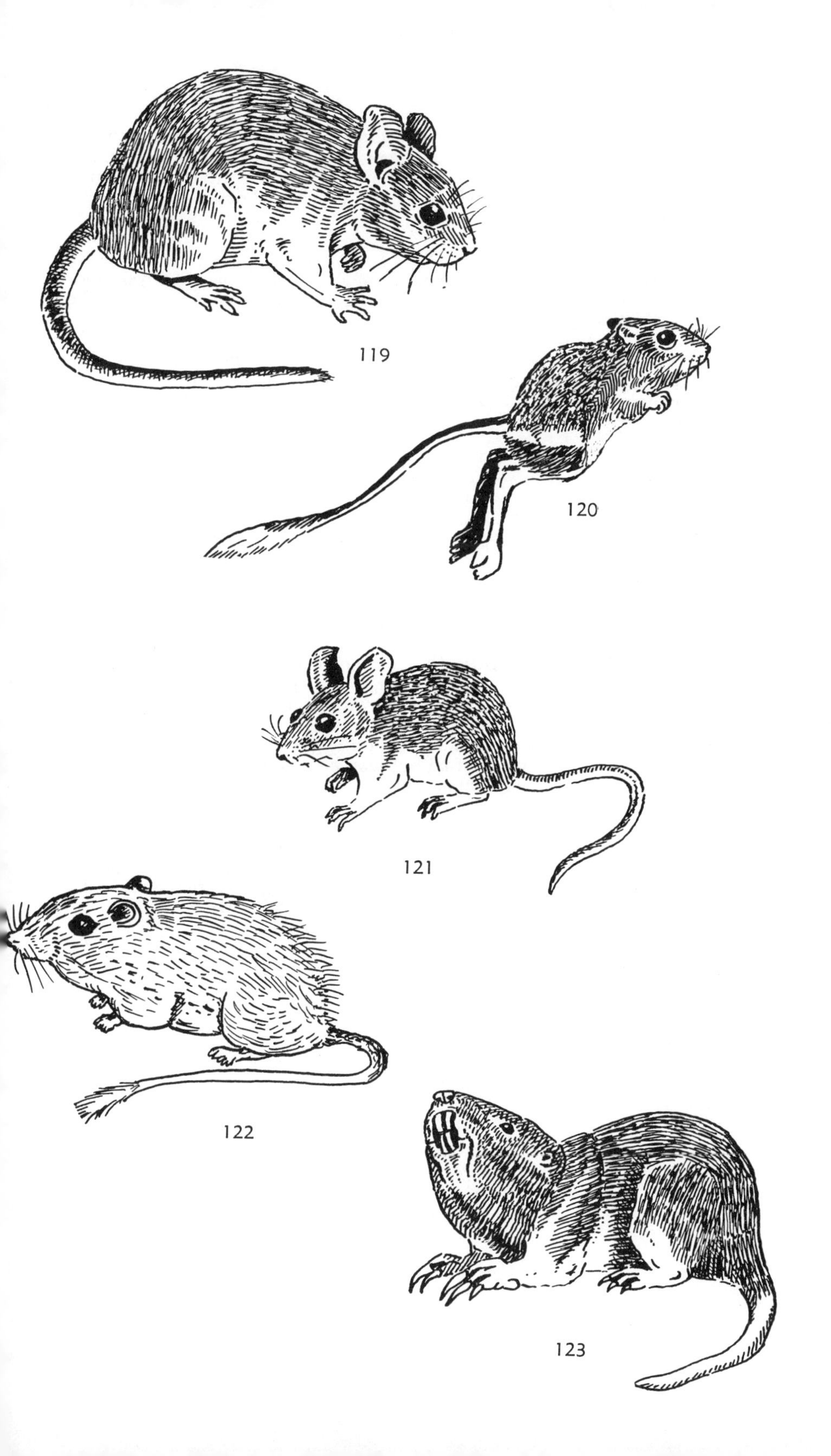
119
120
121
122
123

124 RINGTAIL CAT. *Bassariscus astutus.* 14–16 inches, head and body; 15 inch tail. Shy, short-legged, large-eyed, graceful animal with distinct blackish brown and whitish rings on long tail. Absent from most of the Great Basin Desert. Prefers cliffs and rocky canyons of desert foothills and higher elevations.

125 KIT FOX. *Vulpes velox.* 21–32 inches, including 9–12 inch tail. A small buffy-yellow fox, with big ears, conspicuous black piercing eyes, and handsome black-tipped bushy tail. A gentle, unsuspicious mammal of the open desert plains. Most often seen at dusk.

126 COATI. *Nasua narica.* 20–25 inches, head and body; 20–25 inch tail. Long-tailed, long-snouted, grayish-brown near-relative of the raccoon; distinct white spot above and below eye. Tail often carried erect. Has entered Arizona Upland Desert from Chihuahuan Desert.

127 ARIZONA GRAY FOX. *Urocyon cinereoargentatus.* 20–29 inches, head and body; 11–16 inch tail. A small gray fox with orange or rusty-yellow on sides, throat, and legs. Under parts lighter gray. Like the kit fox it has a black-tipped tail. Tail held out straight when animal is running. Mostly upper Sonoran Life Zone where it inhabits rocky and brushy areas.

128 BOBCAT. *Lynx rufus.* 25–30 inches, head and body; 5 inch tail. Brownish body spotted with darker brown to black. Ear tufts present, but small. Tail black-tipped above. Prefers brushy areas.

129 BADGER. *Taxidea taxus.* 18–20 inches, head and body; 4–5 inch tail. Very broad, heavily built, stout-clawed, short-legged, yellowish-gray animal of the open desert. The face is black, with white stripe along cheeks and from nose backward between ears to shoulders. The feet are blackish. Food: mostly rodents; sometimes, birds, bird eggs, insects, and reptiles. Generally seen alone. Mostly nocturnal but may be seen in daytime, especially in the morning. Rapid, efficient burrower.

130 PECCARY. *Pecari angulatus.* 36 inches long; 20–24 inches high. A distinctly salt-and-pepper-colored, pig-like animal with heavy neck and shoulders. A more or less conspicuous, lighter colored "collar" of bristles on front of shoulders. Often runs in small bands. Prefers desert washes and brushy areas. Enters southwest Arizona, southern New Mexico, and Texas from deserts of Mexico.

124
126
125
127
128
129
130
MVD
MVD

131 HOG-NOSED SKUNK. *Conepatus leuconotus.* 14–18 inches, head and body; 7–12 inch tail. The name refers to the long pig-like snout especially fitted for rooting out grubs. The hair is coarse and, compared with other skunks, short. From southeastern Arizona to New Mexico, southern Texas, Mexico (except Baja California), and Central America.

132 HOODED SKUNK. *Mephitis macroura.* 12–16 inches, head and body; 14–15 inch tail. The belly is always black but the back and tail may be white or the back may be black with two white stripes and the tail black. Between these two color patterns there may be intermediate variants. Called hooded skunk because of the hood or cape of longer, usually white, hairs of the head and neck. Hot valleys of Lower Sonoran Zone of Arizona and New Mexico, southward through Mexico. A very active, nimble skunk with long soft fur.

133 SPOTTED SKUNK. *Spilogale putorius.* 9–12 inches, head and body; 5–8 inch tail. A small skunk with terminal portion of bushy tail white. Body color black, with four longitudinal white stripes on foreparts of body; white stripes and spots on rear half; white spots on face. Nocturnal, fearless little beast. Scent strong but not lasting. Widely distributed on deserts, where it prefers rocky foothills and brushy area.

134 DESERT BIGHORN. *Ovis cremnobates.* 3–3½ feet high and weighing up to 275 pounds. Largest of desert mammals. Distinctive are the massive spiraled horns of the male, and white rump patch. The horns of the female are much smaller. Rough steep terrain of mountains of southwestern deserts. Absent from much of Sonora and most of Chihuahuan area.

131

MVD
132
133
mvd
134

DESERT *Plants*

Strange as it may seem the deserts become, after seasons of favorable rains, America's most beautiful and extensive wild flower gardens—vast seas of color, at once sublime and picturesque. The total number of kinds of plants contributing to this marvelous wild flower exhibition is numbered not in hundreds but thousands, probably between five and six thousand. It is no mean accomplishment to know even half the species; few are the botanists who have knowledge of even the majority of them. A single distinctive desert such as the Mohave, which has received in past ages plant migrants from the north, east, west, and south, may alone boast as many as several hundreds of unique species.

It must at once be apparent that it is impossible to mention and illustrate more than a mere fraction of these flowering plants, perhaps not even one-tenth of them. But an effort has been made to make drawings of the more common, representative, and attractive ones. Since, as yet, travel is much more frequent in the deserts of the United States than in those of Mexico, our own desert plants have received the greater attention.

The desert plants which put on the largest and most startling displays of color are the late winter, spring, and summer annuals, though there is not a season of the year when attractive flowers are not somewhere in bloom. These winter, spring, and summer annuals survive through the seasons of combined heat and dryness in the form of seeds; thus they escape drought rather than withstand it. Mingled with or near these showy ephemerals may be perennial flowering herbs, shrubs, and deep-rooted trees. These withstand seasons and sometimes years of meager rainfall by peculiar adaptations such as inflated stems for water storage, exceedingly deep root systems, wax-coated leaves or leaves of diminutive size which cut down the evaporation possibilities.

Dr. F. W. Wendt and his assistants and botanical colleagues of the California Institute of Technology at Pasadena have carried on through many years patient research on the conditions affecting seed germination and growth of seedling plants, particularly those of the Colorado and Mohave deserts. The following five paragraphs offer a much condensed summary of their extraordinarily interesting and significant discoveries and observations.

Dr. Wendt divides the annual plants into four groups: 1. Those which grow only after summer rains—a limited number of species, most of which have migrated from areas with frequent summer rains such as occur in Mexico. Among these "summer rain opportunists" are the little strong-scented yellow pectis (*Pectis papposa,* 236) and the handsome, colorful fringed amaranth (*Amaranthus fimbriata*). 2. Those which start up after early autumn rains and grow slowly through the winter and are ready to break into flower by springtime. In this category are the great host of best known and plentiful wild flowers such as the many kinds of evening primroses (*Oenothera,* 181, 243), sand verbenas (*Abronia,* 287), and desert sunflowers (*Geraea,* 278), some of the lupines, and the spectacle-pod (*Dithyrea,* 173). 3. Those whose seeds germinate after late autumn or winter rains; among them many of the wild annual buckwheats (*Eriogonum*), gilias, and phacelias. 4. Those with seeds germinating any time of the year, provided there is sufficient moisture; to this group belong several of our wild gourds, the jimson weed, and the Spanish needle (*Palafoxia,* 325).

It has been found that there is a very definite lower and upper limit of precipitation at which the seeds of desert plants will germinate. If the amount of rain is less than 10 mm. it is ineffective for awakening growth of the seed of any plant. If the precipitation is too great, such as occurs at times of cloudbursts, it has a depressing effect and no germination ensues, because certain growth-promoting substances in the seeds are leached out by the large amount of water that passes over them. On the other hand, it must be pointed out that many seeds of wild plants form growth-inhibiting substances which prevent germination, and if these are removed, as by gentle rains, awakening of the embryonic plant within the seed can occur. When there are rains of 15 mm. a limited number of seedlings appear, but not until there are rains of 25 mm. or more does extensive germination take place.

Many seeds of desert plants fall where there are considerable concentrations of salt in the soil; not until rains of sufficient intensity have leached out these deleterious soil salts will the seeds awaken to growth.

Once seedlings appear, the young plants almost always continue to develop until maturity is reached. In some cases, this stage is reached and flowers begin to bloom while the plant is still exceedingly small and has but one or, at most, two or three leaves. These are among the flowering dwarfs we so often see during seasons when late rains do not materialize.

Some of the desert shrubs apparently have a number of ways of preventing competition from other plants. Ordinarily only one new bush may grow where another has died. Many seeds may germinate but few of the seedlings have the merest chance of growing to maturity. The old established bushes may actually "kill off" the young competitors in the immediate vicinity, even though these are their own offspring. It has been found that the roots of a considerable number of plants, such as burro bush, thamnosma, guayule, and creosote bush, actually excrete into the soil inhibitory toxic substances which under certain conditions may repress the root growth of the young plants. They may start to grow but because of the inhibitor soon shrivel, then die. This is one explanation for the wide regular spacing of many of our desert shrubs. These toxic materials affect only certain plants; others are toxin tolerant.

There are certain of the desert trees and shrubs which grow only in the bottoms of sandy washes. This is not wholly due to their water needs but also results from the peculiar germination requirements of their seeds. Ironwood, smoke tree, and certain palo verdes (*Cercidium aculeatum*), for instance, have exceedingly hard-coated seeds. The embryonic plant within the seed cannot possibly break through unless this indurated covering is worn thin by the grinding action of moving sand and gravel. This is accomplished at the time of heavy rains or flash floods, when the seeds are carried along in the rushing sand- and gravel-laden streams.

There is a considerable group of desert annuals, among them some of the wild buckwheats, which come up late in the season of rains and continue growth during the ensuing dry warm days of very late spring and early summer. They bloom for some time after growth in stature has ceased. When such plants are pulled from the dry sand they are seen to have not only a very limited root system but to be growing in soil that appears to furnish them little or no water. It is believed that these plants flourish in spite of the apparent soil dryness because they are able to take up water vapor from the air at night, then conserve and utilize this water during the day. On cloudless nights of early summer the relative humidity may be sufficiently high to make this absorption of water possible.

Absorption of water from saline and alkaline soils is relatively difficult for most desert plants. If the concentration of the soil solution becomes high, water is withdrawn from the roots, causing a kind of permanent wilting, a stunting of growth, or actual death. Certain species, such as greasewood (*Sarcobatus*), glasswort (*Salicornia*), and ink weed (*Sueda*), and many kinds of salt bushes (*Atriplex*), are believed to be able to flourish in such alkali and salt soils because they are genetically conditioned to continuously maintain a higher osmotic pressure in the root cells than obtains in the water among the soil particles surrounding the rootlets.

The salt and alkali plants are known as halophytes. They grow most frequently on the borders of dry lakes and around seeps and small streams of volcanic hills and mesas. It is always interesting to note the zonation of such plants about dry lakes. The most salt-tolerant species, such as greasewood and glasswort, form an inner zone with concentric circles of less and less salt-tolerant ones toward the outside. The usual coloration of halophytes is gray. They resemble true xerophytes in having densely hairy leaves (hence the gray color), reduced leaf surface, etc., because they suffer from a kind of physiological drought.

Some species of the deserts' annual wild flowers, such as certain phacelias and crucifers, are almost always associated with certain shrubs, seldom occurring in the bare spaces between. A most favorable condition promoting the growth of such shrub-dependent annuals is the accumulation about the shrub bases of humus-rich mounds of windblown soil and organic debris consisting of dead leaves, broken plant stems, and similar materials. These debris heaps are especially prevalent under the more twiggy shrubs. Annuals such as the wild forget-me-nots (*Cryptanthe*) are quite independent of shrubs and occur in greatest abundance in the sunlit interspaces.

The line drawings of trees, shrubs, perennial herbs, and often colorful annual wild flowers which I have made have been arranged according to the predominant flower colors. This is a simple classification that will appeal to the many not acquainted with technical keys.

By often examining the illustrations and carefully reading the text material accompanying them, the plants *most common and characteristic of each region* will become well known and will soon be recognized in the field. I have given particular attention to illustrating the important plants mentioned in the chapters dealing with each of the major desert areas.

Try, as you travel, to get acquainted first with these outstanding and conspicuous kinds. Because trees and shrubs are so readily seen, and are such prominent members of the flora throughout the entire year, special attention has been given to illustrating these. The author is well aware that because of the great number of plants he has not drawn, there will be moments of frustration when a plant cannot be identified. He suggests that instead of being discouraged, one should seek those that are illustrated. Most of the drawings, except of course those of the cacti and some of the larger trees and shrubs, show the specimens reduced to slightly more than one-third the usual size.

135 BIG SANDBURR. *Cenchrus palmeri.* This is a very abundant sandburr in sandy areas of Sonora and middle parts of the Colorado Desert in Mexico. It is both a summer and a winter ephemeral, appearing in great abundance after some of the erratic heavy rains which visit these regions. The rather handsome burrs, supporting numerous sharp bristles, are reddish brown when mature and very obnoxious.

136 BIG GALLETA. *Hilaria rigida.* From deserts in southern Utah and Nevada and on south to the Colorado Desert and the plains of Sonora, this coarse-stemmed perennial grass flourishes on alkaline-free, gravelly soils. It is a valuable source of food for cattle when found in places where water is available. The stems are covered with a felty mass of soft hairs.

137 TOBOSA GRASS. *Hilaria mutica.* A well-known grass of the arid lands of Texas, New Mexico, Arizona, and the northern Chihuahuan Desert. The culms do not have the felt-like covering that those of BIG GALLETA grass have.

138 RED MOLLY. *Kochia americana.* Perennial herb with numerous erect branches from a woody base. The narrow leaves, oval in cross-section, are grayish-green. A plant of alkaline flats of the northern Mohave and Great Basin deserts, to New Mexico.

139 TRUE GREASEWOOD or CATERPILLAR GREASEWOOD. *Sarcobatus vermiculatus.* Widely distributed and conspicuous shrub of the alkaline flats of Utah, Nevada, and the Death Valley region in California. Often associated with shadscale and inkweed around the borders of dry lakes.

140 CHEESE BUSH. *Hymenoclea salsola.* Low, much-branched, bright green shrub of cheesy odor; attractive because of its conspicuous, spirally arranged, scarious-winged fruits. Sandy washes and occasionally dry desert slopes of southern Utah, Arizona, southeastern California, and Baja California.

141 CANDELILLA. *Pedilanthus macrocarpus.* A wide-spreading, low, erect-stemmed shrub, with thick, milky, very sticky sap. Its fleshy stems do not freely branch and spread outward as do those of JATROPHA, its near spurge relative. Baja California. The nearly related *Pedilanthus pavonis* has a small scarlet fruit with horn-like projections.

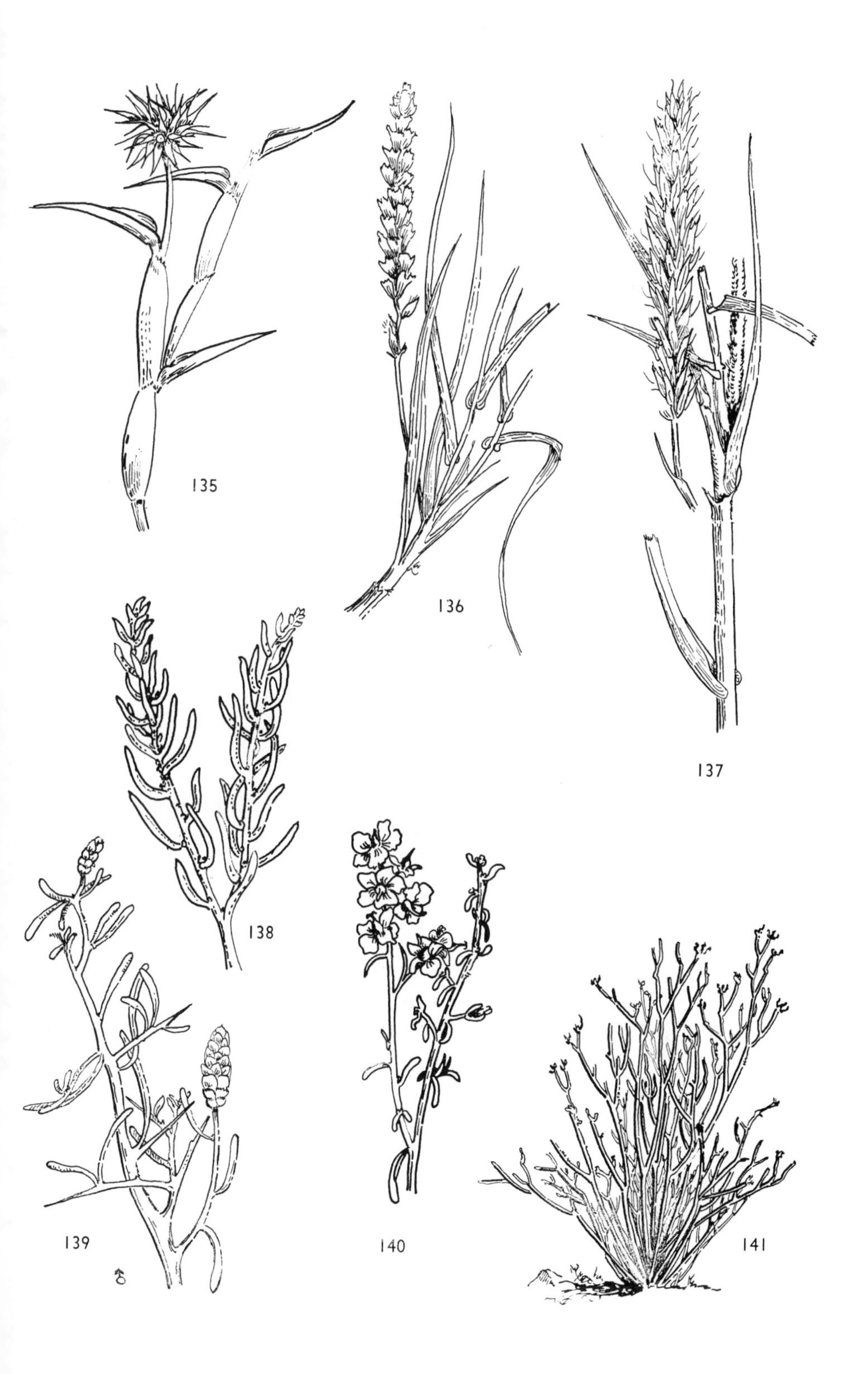
135
136
137
138
139
140
141

142 HOTE TREE or JITO. *Forchammeria watsoni.* A densely leaved, dark-green tree belonging to the caper family. There are few dead limbs or twigs, a fact suggesting its hardiness. Open deserts of Sonora and Baja California, along with ironwood, palo verde, and senita cactus. The compact rounded crown is often flat on top. Flowers in March and April. The wood is lightweight and spongy.

143 JUMPING BEAN. *Sapium biloculare.* Shrub to small tree with very toxic milky sap. The rounded larva-infested seeds "jump" much after the manner of the MEXICAN JUMPING BEAN (*Sesbania pavoniana*). Sapium is a member of the spurge family. Southwestern Arizona, Sonora, and deserts of northeastern Baja California.

144 GRAY CRUCILLO. *Condalia lycioides canescens.* Thorny, gray-barked spreading shrub up to 9 feet high. Salton Sea Basin, eastward to southern Arizona, south to Sonora and Baja California. The fruits are eaten avidly by quail and several other birds.

145 SPINY SALT SAGE or BUD SAGEBRUSH. *Artemisia spinescens.* Low, compact, spiny shrub with stout woody stems and small, round, green flower heads. It is one of the dominant plants in sandy and semi-alkaline flats of the middle Mohave Desert and northward in the Great Basin Desert as far as Wyoming.

146 LEATHERPLANT. *Jatropha spathulata.* One of the oddities among small shrubs one sees in desert areas of the Chihuahuan Desert, Sonora's plains, and southern Arizona. The leathery spatula-shaped leaves spring from the thick, flexible, reddish-brown stems after spring and summer rains. The bark is used for tanning and dyeing. The flowers are very small.

147 JUNCO or CRUCIFIXION THORN. *Koeberlinia spinosa.* Rigid spiny branched shrub of the intermountain plains of the Chihuahuan Desert and desert grasslands of southeastern Arizona. The fruit is a shiny black berry. The small flowers occur in compact bundles.

148 MEXICAN CRUCILLO. *Condalia spathulata.* Dense, twiggy shrub up to 5 feet high with numerous small, sweet, edible berries. Eastern California, southern Arizona to Texas; Chihuahua and Baja California in Mexico.

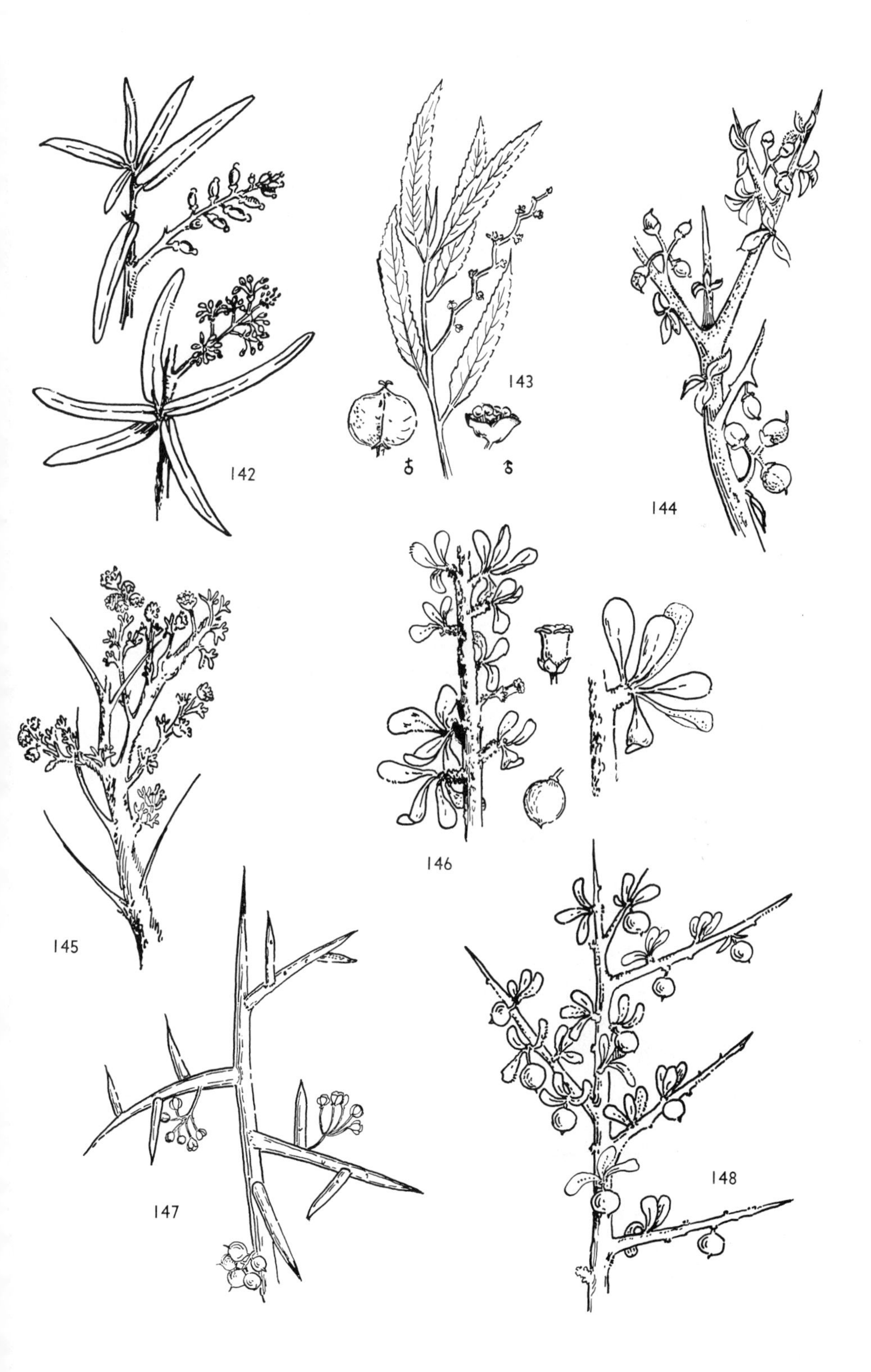

142
143
144
145
146
147
148

149 ORGAN PIPE CACTUS or PITAHAYA DULCE. *Lemaireocereus thurberi.* Stout-stemmed cactus branching from the base and growing up to 20 feet high. The fruits are prized for their sweetness. The flowers open at night. Southwestern Arizona to Baja California and Sonora.

150 DESERT POINSETTIA. *Euphorbia eriantha.* Freely branching herb attaining its best growth in the shelter of shrubs. The true leaves and involucral leaves surrounding the small "flowers" are not scarlet but bronzy green. Low areas in Colorado Desert, Sonora, and Baja California.

151 SHADSCALE. *Atriplex confertifolia.* This low, woody, spiny shrub of rounded form is, along with sagebrush, a predominant plant over much of the Great Basin Desert and higher parts of the Mohave Desert. It is very tolerant of the high alkaline-content, heavy soils; it often forms almost pure stands over great areas.

152 WINTER FAT. *Eurotia lanata.* A low bush with erect herbaceous stems, 1 to 2½ feet high, and particularly conspicuous because of the gray-white foliage and long-haired fruits. A most valuable winter grazing plant of the Mohave and Great Basin deserts.

153 DESERT HOLLY. *Atriplex hymenolytra.* A handsome low silver-green salt bush of the alkaline soils of hot washes and hillsides. The leaves sometimes turn reddish. Southern Arizona, southern Nevada, and Mohave and Colorado deserts of California.

154 ARROW SCALE. *Atriplex phyllostegia.* One of the oddest of the salt bushes. The leaves are semi-succulent. Found in "clan-like groups" in soils permeated with alkalies. Often seen in areas of salt grass and rabbit brush. Western and middle Mohave Desert.

155 TRUE SAGEBRUSH or THREE-TOOTH SAGE. *Artemisia tridentata.* Gray-green shrub occurring widespread over high deserts from northern Baja California to Nevada, Utah, western Oregon, and Idaho. Note the three teeth at the ends of the leaves.

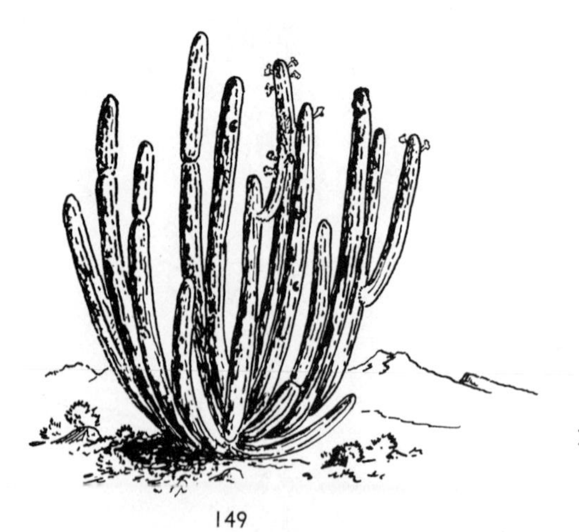

149

DESERT PLANTS (*greenish white*)

150
151
152
153
154
155

156 NOLINA. *Nolina parryi.* Only superficially does nolina resemble a yucca. Instead of the stiff dagger-like leaves, large flowers, and flat seeds of the yucca, it has thin pliable leaves, small flowers, and small round seeds surrounded by papery, dry, fawn-colored wings. Many think that the dry plumes of ripe fruits of the nolina are more beautiful than the large flower-panicles of the yucca. This handsome plant is common in the Joshua Tree National Monument. A smaller nolina but with many characteristics of the above is BIGELOW'S NOLINA (*Nolina bigelovii*), largely confined to the drier hills and mountains of the Colorado Desert, southern Arizona, San Diego County in California, and northern Baja California. Bigelow's Nolina is also found in the Joshua Tree National Monument.

157 BAILEY'S YUCCA. *Yucca baileyi.* A handsome large-flowered yucca seen in the juniper areas of parts of the Painted Desert. The leaves are narrow in proportion to their length. The large capsules open when ripe, allowing the seeds to fall. The specimen shown here was found near Winslow, Arizona.

158 CARDON. *Pachycereus pringlei.* Largest of all desert cacti. Plants may grow to 50 or 60 feet high and weigh several tons. Sonora and Baja California.

159 SAHUARO. *Cereus giganteus.* Arizona state flower, the most massive cactus in the United States. It is exceeded in size only by the large *Pachycereus* of Baja California and some of the giant cacti of southern Mexico and Brazil. The beautiful white flowers are borne in crownlike clusters at the ends of the stout branches. Southeastern Arizona, Sonora, and in a few places along the Colorado River in southeastern California.

160 CARDON. *Pachycereus calvus.* One of the common giant cacti of southern Baja California's Arid-Tropical Zone. Named *calvus* (Latin, "bare, bald") because the tops of the stout branches are often devoid of spines. Some authorities think it is but a form of *Pachycereus pringlei.*

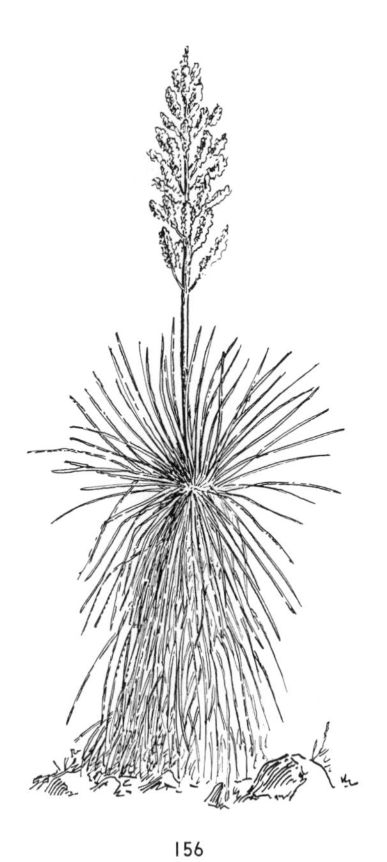

156

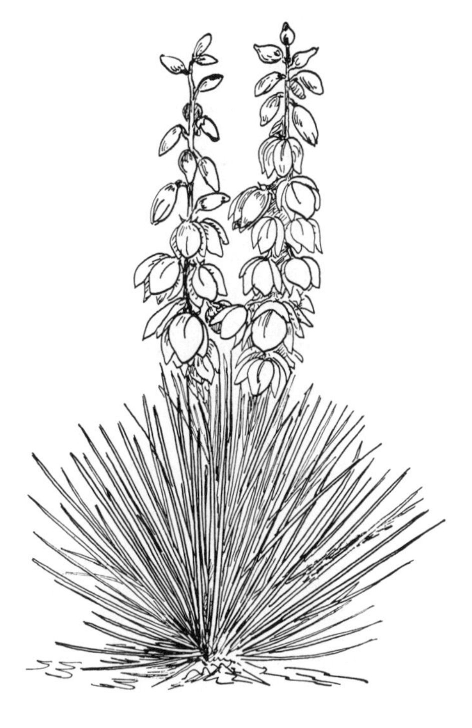

157

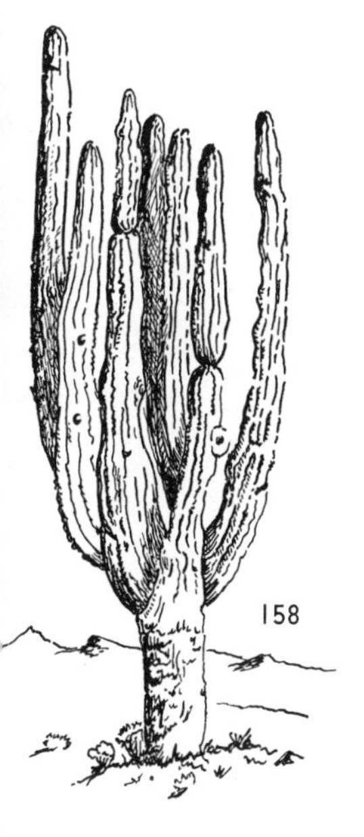

158

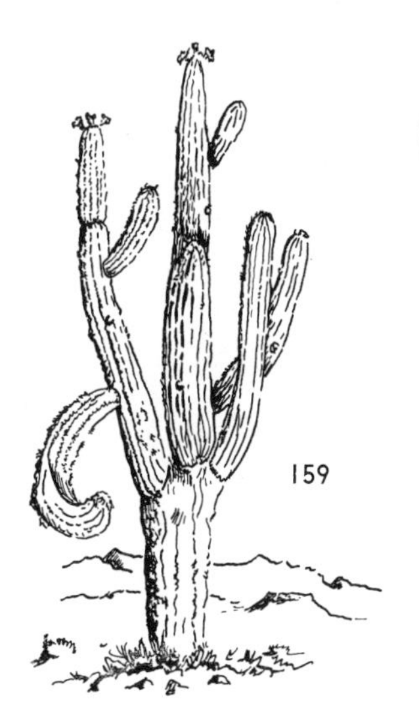

159

160

161 PALMILLA or SOAP WEED. *Yucca elata.* May be recognized by its well-developed, trunk-like stem; long flexible green, white-margined leaves having thready edges; and tall flowering stalk with large ellipsoid panicles of white or cream flowers. These usually appear about the first of June to the middle of July. This handsome yucca occurs in a wide belt from Pecos County, Texas, northwest and west through southern New Mexico, south-central Arizona, and northern Mexico. Indians roast the lower flower stalk and lower portion of the stem for food.

162 TORREY YUCCA. *Yucca torreyi.* Single- to several-stemmed yucca of southwestern Texas, southeastern New Mexico, and northern Chihuahua and Coahuila in Mexico. The long leaves are margined with rather coarse curly fibers. The fruits are fleshy.

163 CRUCIFIXION THORN. *Canotia holacantha.* Shrub to small tree, 6 to 18 feet high; the leafless, spine-tipped rigid branches occur in somewhat broom-like clusters. It occurs singly or in thickets of the upper rocky slopes of the northern part of the Arizona deserts to eastern Mohave Desert.

164 SOTOL. *Dasylirion wheeleri.* This sotol has numerous strap-shaped, gray-green leaves with spiny margins, the prickles directed forward. The flower shaft may be up to 15 feet high. Chihuahua and western Texas to Arizona. Other species of sotol are found in the same area. The SMOOTH-LEAVED SOTOL (*Dasylirion leiophyllum*) is found in the Texas desert.

165 MOHAVE YUCCA. *Yucca schidigera.* This yucca is common over much of the Colorado and Mohave deserts, often dominating wide upland slopes and basins. Distinctive are its long, stiff, yellowish green leaves with splinter-like marginal fibers. The leaf fibers were utilized in many ways by the Indians, for rope making, cloth weaving, the manufacture of moccasins, etc.

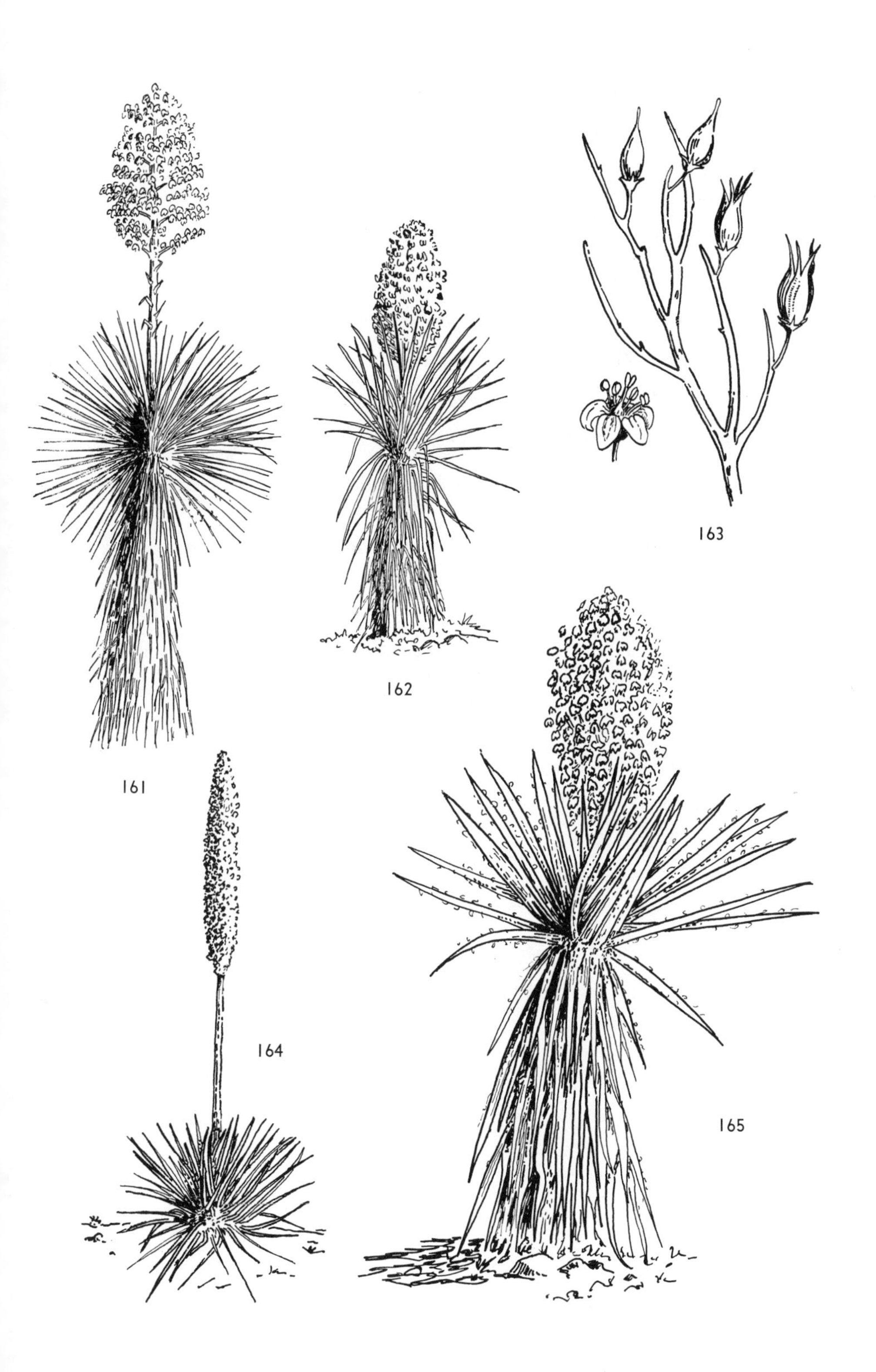
161
162
163
164
165

166 SANDPAPER PLANT. *Petalonyx thurberi.* There are several kinds of sandpaper bushes found on our deserts, all with small sweet-scented flowers and rough sandpapery foliage and stems. Most common in sandy soil.

167 LIPPIA. *Lippia wrightii.* Branching shrub about 3 feet high with slender stems, strongly scented leaves, and spikes of very fragrant, tubular, tiny white flowers. A plant of rocky hills of the deserts of western Texas, New Mexico, southern Arizona, and adjacent Mexico in Sonora, Chihuahua to Durango.

168 SPINY MENDORA. *Mendora spinescens.* A low, spiny, perennial shrub with showy white flowers and peculiar double fruits of greenish color. Eastern Mohave Desert of California and southern Nevada.

169 DESERT ZINNIA. *Crassina pumila.* Low perennial bush with flowers sometimes pale yellow, sometimes white. Arid west Texas to southern Arizona and northern Mexico. Prefers limestone soils.

170 PRICKLY POPPY. *Argemone hispida.* Widespread in flat sandy deserts, often forming beautiful roadside borders, especially in Sonora, Mexico. The yellow-centered, white-petaled flowers are fragrant. An orange-yellow milk exudes from the injured stems.

171 THORN APPLE. *Datura meteloides.* This is often a desert roadside weed but very attractive because of its large, pale-lavender, sweet-odored flowers, which open during the night. All parts contain a powerful narcotic. There are five teeth on the edge of the corolla. The seeds of the spiny fruits are brown when ripe. A somewhat similar plant is *Datura discolor,* with shorter flower-funnel and with ten teeth on the corolla edge. Its leaves are green instead of gray-green and the seeds black instead of brown.

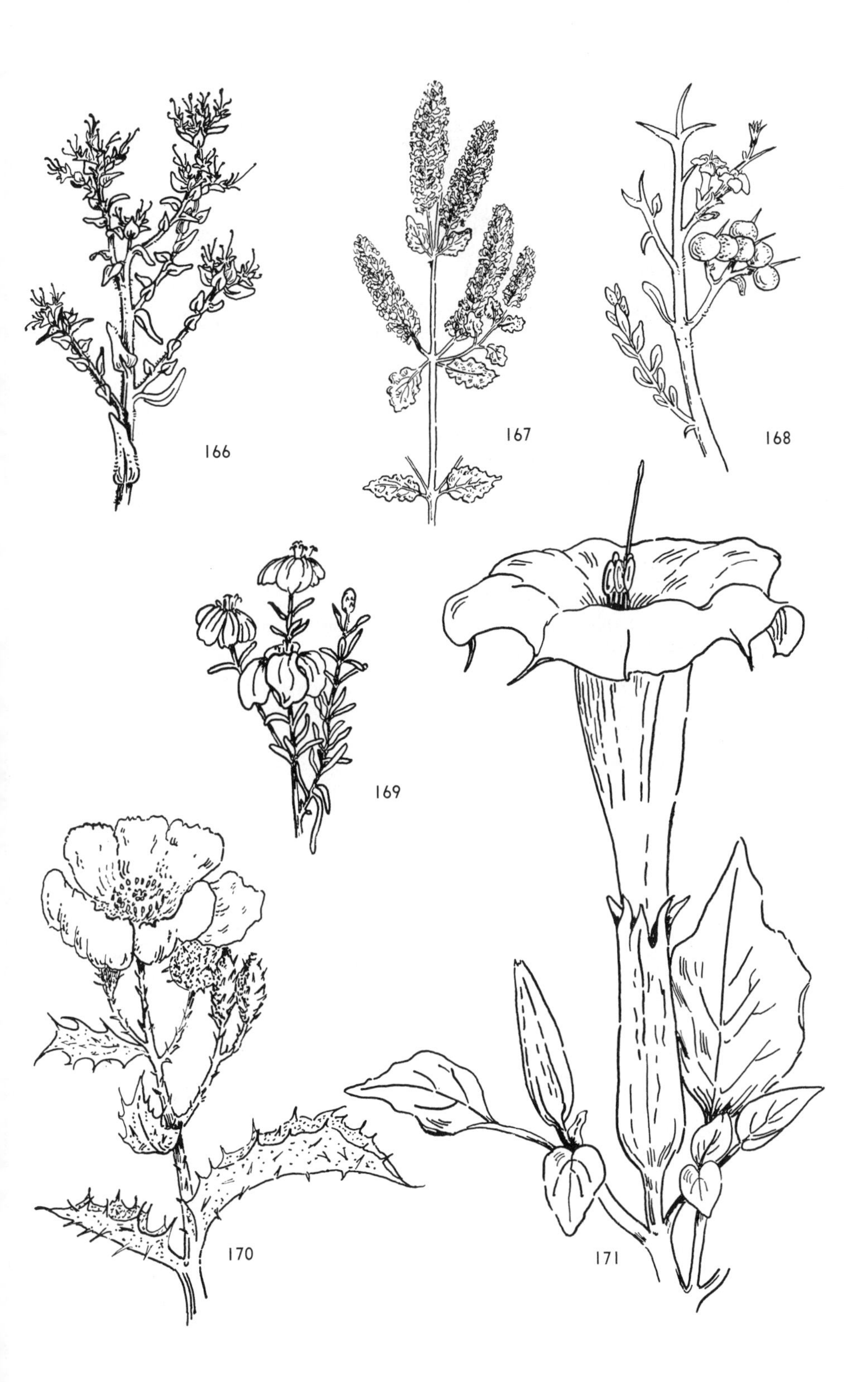
166
167
168
169
170
171

172 WISHBONE BUSH. *Mirabilis bigelovii retrorsa.* Bright green, tender perennial herb with white flowers opening in the evening. Called "wishbone bush" because of its manner of branching. Gravelly flats and rocky surfaces of Colorado and Mohave deserts.

173 SPECTACLE POD. *Dithyrea californica.* A sweet-flowered crucifer of the low sandy deserts; certain to attract attention because of its double fruits. Flowers white to purplish. Southern Nevada, western Arizona, southeastern California, and Baja California.

174 PLANTAIN. *Plantago sp.* Small spring or summer annual hairy herbs, 3 to 4 inches high. Leaves all basal. Flowers small and borne in elongated spikes. These are plants of low flat places with fine loamy or sandy soils. There are many species. Common to all deserts.

175 BRANDEGEA. *Brandegea bigelovii.* A perennial vine with large water-storing roots and delicate green leaves. It grows about the bases of large desert shrubs and trees, often climbing up among the branches. The white flowers are very small, as are also the spiny fruits. There is great variation in the shape of the tender leaves. Southwestern Arizona, southeastern California, and northwestern Mexico.

176 LANCE-LEAVED DITAXIS. *Ditaxis lanceolata.* A silvery green, low-growing, perennial spurge of low rocky mountainsides. The small flowers are faint greenish white. Desert bighorn browse on it whenever they find it. It is one of the spurges found in the Colorado Desert eastward to southern Arizona and northern Baja California.

177 PARRY GILIA. *Gilia parryae.* A much-branched small annual with flowers very large for its size. A vigorous daytime bloomer with white to purplish-white flowers. Sandy soils.

178 PUNCTURED BRACT. *Oxytheca perfoliata.* One of the strangest of the desert's low colorful herbs, especially handsome when the saucer-like bracts turn reddish. It sometimes forms mats almost a foot across. Especially abundant in the Mohave Desert.

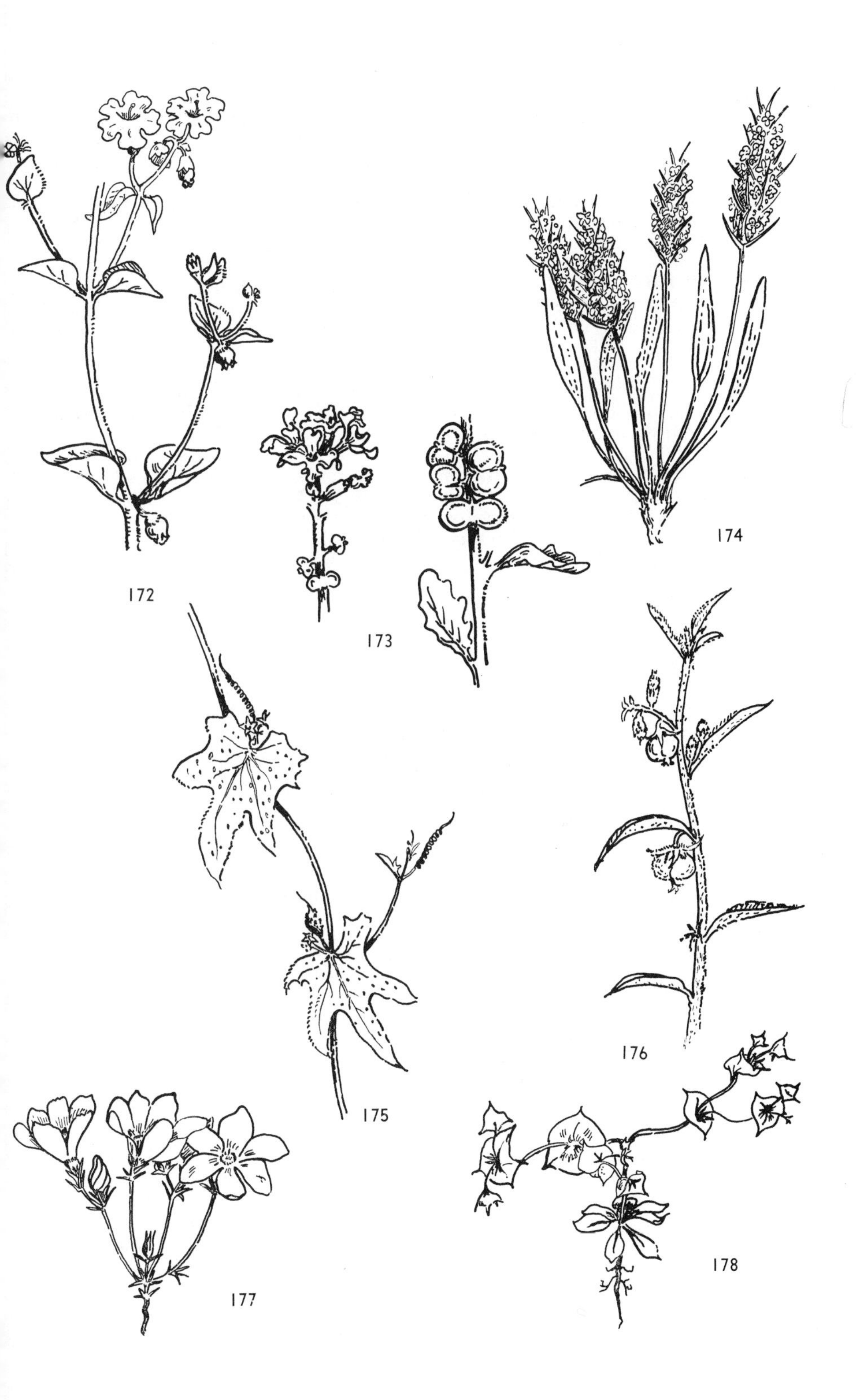
172
173
174
175
176
177
178

179 DESERT ALYSSUM. *Lepidium fremontii.* A rounded, bushy, green-stemmed perennial 8 to 20 inches high, densely covered in late spring with myriads of small sweet-smelling flowers. Sand washes and rocky creosote plains of the Mohave Desert to Nevada, southern Utah, and Arizona.

180 DESERT TOBACCO. *Nicotiana trigonophylla.* An ill-smelling but handsome biennial or perennial herb. Both California deserts to Texas and Mexico. Used by the Yuma and other Indians for smoking.

181 WOODY BOTTLE-WASHER. *Oenothera decorticans.* The dried plants, with woody core and splitting capsules, are certain to attract attention. The green plants have white, shining bark which splits off. The flowers are white to pink.

182 VASEY'S WAND SAGE. *Salvia vaseyi.* Named for George Vasey (1822–93), well-known authority on grasses. Canyons and hillsides bordering the northwestern part of the Colorado Desert.

183 SONORAN CAPER. *Atamisquea emarginata.* Densely branched, ill-scented leafy shrub 3 to 18 feet high. A relative of BLADDER POD (*Isomeris*). The leaves are green on the upper surface, scaly beneath. The sweet-scented, four-petaled flowers are small. Drawing shows one mature fruit capsule and five immature ones. Dry hillsides, wash borders, and plains; Sonora and Baja California.

180
181
182
183

184 THREADLEAF WORMWOOD. *Artemisia filifolia.* Silver-green shrub up to 3 feet high, very leafy; its half-inch-long leaves almost threadlike. The small yellow flower-heads are found in long, densely leaved panicles. The Mexicans of Chihuahua call the plant "estafiate" and use a tea made from it as an intestinal worm remedy and for stomach ailments. Widespread in sandy areas from Wyoming to Texas, northern Mexico, and Nevada.

185 SAND BLAZING-STAR. *Mentzelia involucrata.* Flowers pale cream with crimson-tinged center. An annual with several stout silvery barked stems 4 to 16 inches high; leaves covered with stiff hairs. Sandy areas and low hot washes and canyons of the low warm deserts of southeastern California, western Arizona, and Sonora.

186 EMORY ROCK DAISY. *Perityle emoryi.* Desert annual with brilliant white flowers with yellow center. Especially frequent on rocky hill and canyon sides. Foliage lively deep green. Both California deserts, southwestern Arizona, and adjacent Mexico.

187 DESERT STAR. *Eremiastrum bellioides.* Dwarf winter annual with yellow disk and white to rose-tinged rays. Often occurs in dense aggregations in sandy places. Southern Utah to Sonora, southeastern California.

188 FREMONT PINCUSHION. *Chaenactis fremontii.* A simple, erect or branching annual from 4 to 16 inches high; flowers white. Frequent under and about creosote bushes. Southwestern Utah, western Nevada, western Arizona, and southeastern California.

189 PEBBLE PINCUSHION. *Chaenactis carphoclina.* An inhabitant of dry rock-floor or patina flats. Flowers near-white. Both California deserts to New Mexico and Mexico.

190 DESERT CHICORY. *Rafinesquia neo-mexicana.* Weak-stemmed annual generally growing in the shade or up through the branches of twiggy shrubs. The ray flowers have rose-colored veins on the under side. Western Texas to southern Utah, southward through Colorado Desert to Baja California.

184

DESERT PLANTS (*white*)

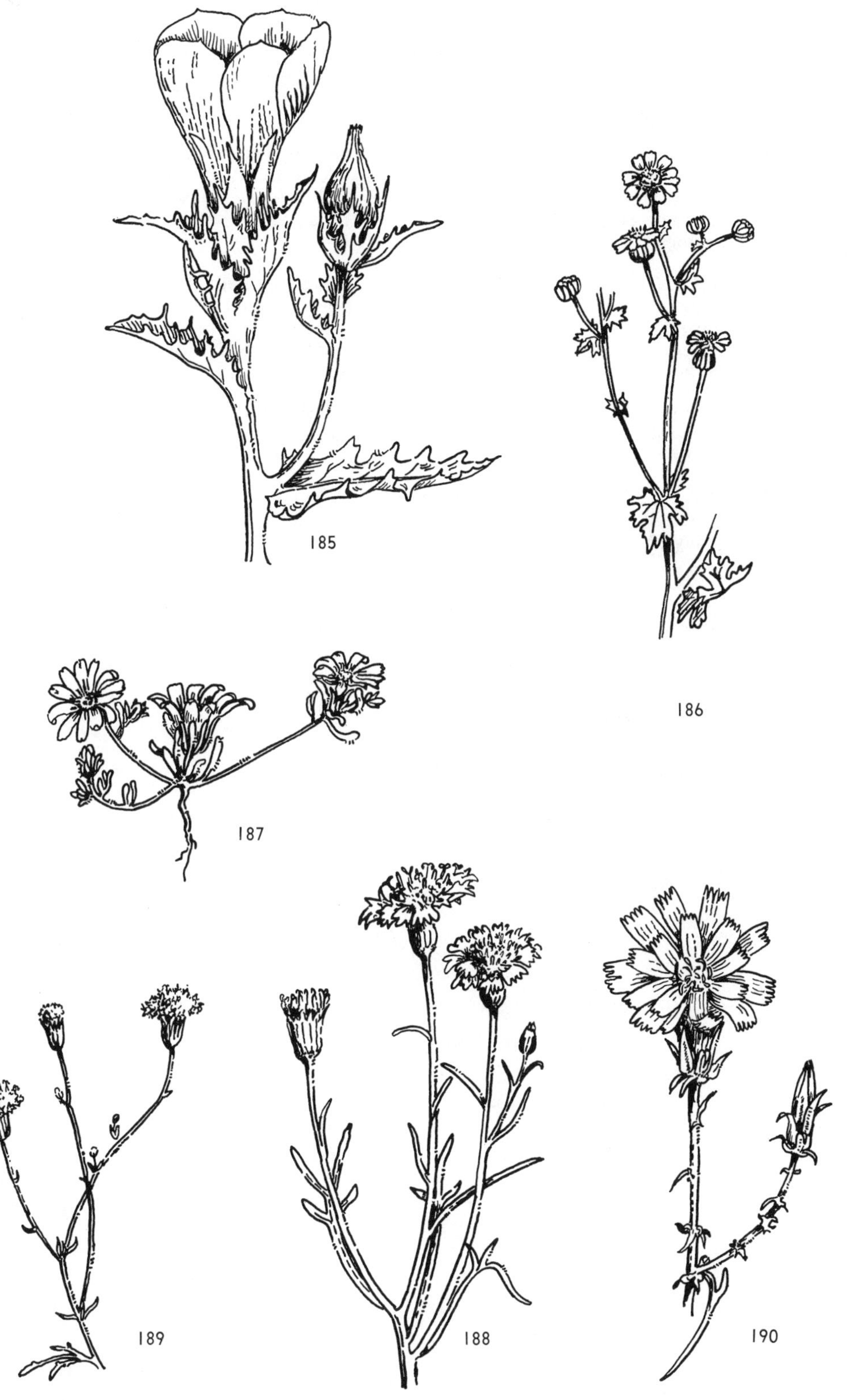
185
186
187
189
188
190

191 PALO BLANCO. *Lysiloma candida.* Tree, sometimes up to 23 feet high, often with several trunks and smooth white bark. The bark is so often sought for tanning that the trees have suffered much. Rocky slopes and canyons of southern Baja California.

192 VARNISH-LEAF ACACIA. *Acacia vernicosa.* Very spiny, low shrub with numerous very small, very viscid bipinnate leaves. Although so small, goats feed on them. Flowers in small rounded whitish "heads" with purple tinge. Common on the plains of Chihuahua to Zacatecas; west Texas and southern Arizona.

193 NIGHT-BLOOMING CEREUS. *Cereus greggii.* Slender, elongate, branched, prominently ribbed cactus springing from a large carrot-shaped root. The strongly scented, fragrant flowers last but one night. Fruits dull-red when ripe. The lead-colored stems often grow upward within the numerous branches of shrubs and are then quite hidden.

194 PAPER-POD MIMOSA. *Mimosa laxiflora.* Shrub with few short spines and pale-green leaflets. The flowers occur in white spikes. The papery pods are light tan.

195 COVILLE'S ACUAN. *Acuan covillei.* A woody-stemmed shrub with numerous greenish-white, dense flower heads and long, flat, reddish-brown, many-seeded pods. Northern Sonora in vicinity of Hermosillo.

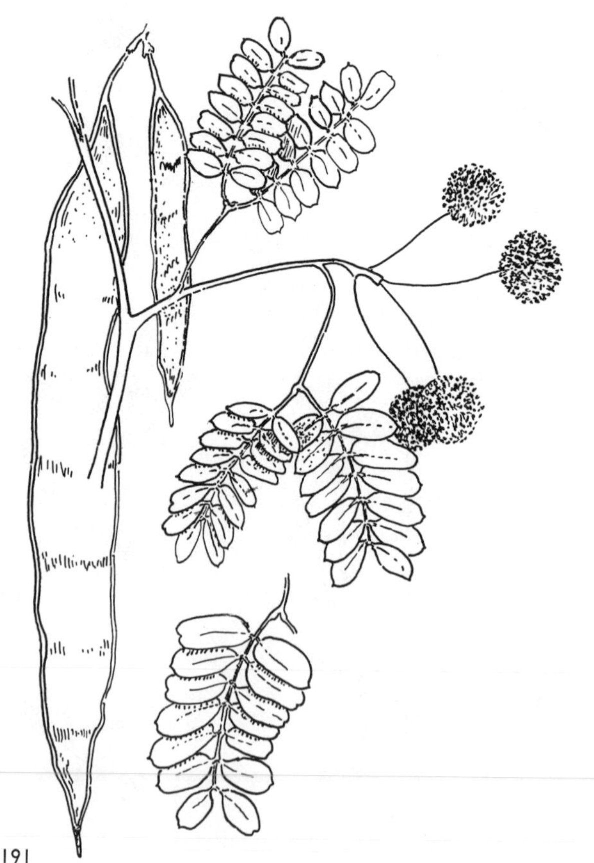

DESERT PLANTS (*white*)

191

192

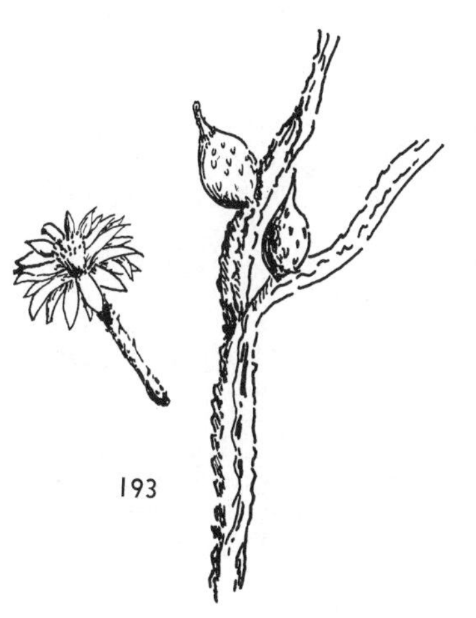

193

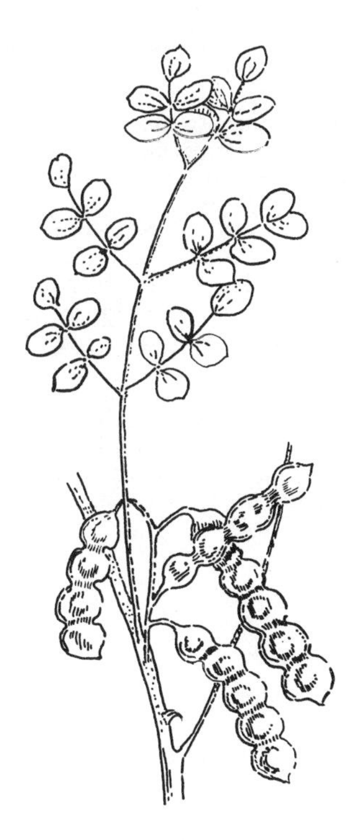

194

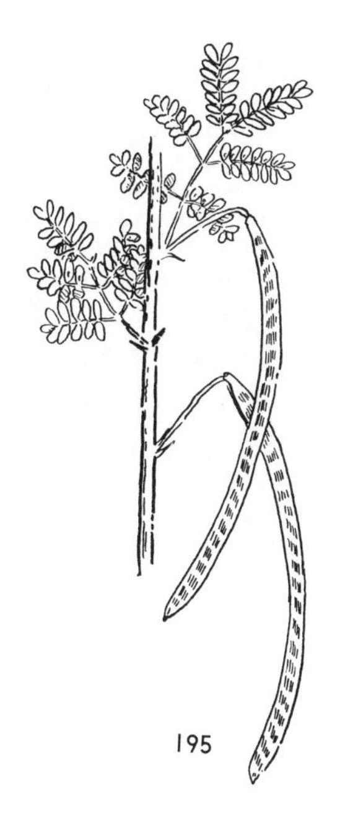

195

196 NELSON AGAVE. *Agave nelsoni.* More than 175 species of century plants are found in Mexico and the southwestern United States. This one is common on rocky areas of north-central Baja California. The gray-green leaves bear detachable marginal teeth; the spine at the end is straight. Plants' associates are cardon, idria, and Pitahaya agria.

197 LECHUGUILLA. *Agave lechuguilla.* Abundant on dry limestone mesas and cliffs of the northern Chihuahuan Desert (southwestern Texas, southeastern New Mexico, and northern Mexico). It often occurs in colonies.

198 PANCAKE CACTUS. *Opuntia chlorotica.* A cactus of higher elevations; of tree-like form sometimes up to 8 feet high, and formed of large disk-like joints 3 to 8 inches in diameter. The spines are close-set and yellow. Flowers 2 to 3 inches wide; fruit purple and edible. Both California deserts to Nevada, Arizona, Baja California, and Sonora.

199 BIGELOW CHOLLA or JUMPING CHOLLA. *Opuntia bigelovii.* A handsome widespread species of the broad, well-drained, rocky detrital fans issuing from the mouths of large canyons. It is propagated only by the detached joints, the seeds being sterile. A favorite nesting site for the cactus wren. Southwestern Utah, southern Nevada, to Arizona, Sonora, and Baja California.

200 BARREL CACTUS or BISNAGA. *Eichinocactus acanthodes, E. wislizeni, E. lecontei.* Several barrel cacti are found upon our deserts, Leconte's and Wislizenius' specimens being among the largest and tallest. The flowers are borne in a circle near the crown. All have stout, straight or curved, hooked spines. After rains the juice extracted by crushing the pulp is potable. The specimen illustrated is the common Colorado Desert barrel cactus (*E. acanthodes*), with straight to hooked spines.

201 NIGGERHEADS. *Echinocactus polycephalus.* Many-stemmed, forming large rounded mounds. The flowers are yellow, the fruit densely woolly. Mohave Desert to Utah and Arizona; generally in rocky areas.

202 TASAJILLA. *Opuntia leptocaulis.* Compact, often bushy cholla with short slender end-joints usually at right angles to the branches. Flowers small, greenish or yellowish; fruits smooth, red, rarely yellow. This long-spined cactus is abundant in the northern Chihuahuan Desert and often forms dense thickets; found also in desert areas of Texas and New Mexico.

203 DARNING-NEEDLE CACTUS. *Opuntia ramossisima.* A much-branched shrubby species with long yellow, slender spines which are set in the center of diamond-shaped areas adorning the surface of the slender branches. The flowers are unattractive, but the fruits are very handsome because of the numerous slender, silvery spines. Both California deserts to Nevada, Arizona, and Sonora.

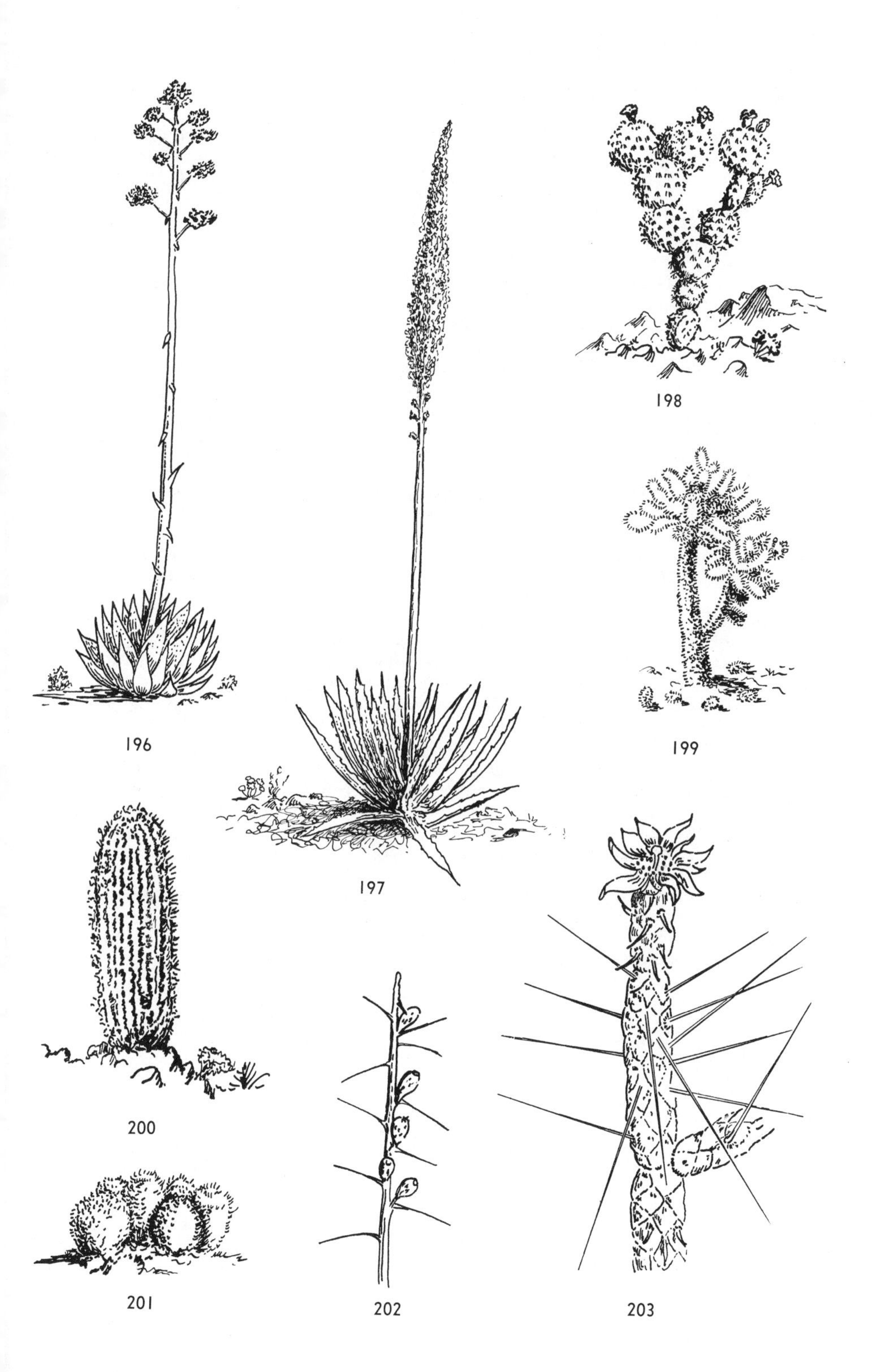

196
197
198
199
200
201
202
203

204 PRINCE'S PLUME. *Stanleya pinnata.* Perennial, many-stemmed plants with tall showy spikes of "spidery-like" yellow flowers. They prefer seleniferous soil and when growing in it absorb the poison. Indians used the leaves for greens, removing the poison by pouring off the boiling water, then recooking. Widespread from North Dakota to Idaho, south to Texas, Arizona, and Mohave Desert. *Stanleya elata* has yellow to almost-white flowers, and generally entire (i.e., not divided) leaves. It is a plant of the northern Mohave Desert and northern Arizona.

205 YELLOW PEPPERGRASS. *Lepidium flavum.* Low, yellowish-green, very early annual of the Mohave Desert plains. Usually found under or near protecting shrubs. A spicy salad plant.

206 DESERT MISTLETOE. *Phoradendron californicum.* This is a leafless mistletoe growing for the most part in clusters in leguminous trees such as palo verde, ironwood, and mesquite. The pearly white to coral-pink berries are an important source of food and water for a number of desert birds. The very small male flowers are exceedingly fragrant.

207 PYGMY GOLD POPPY. *Eschscholtzia minutiflora.* A very small-flowered poppy with leaves well distributed along the stems. Found from the Mohave Desert to northern Baja California and western Arizona. Another small-flowered poppy with finely dissected leaves crowded at the base and with deeply pitted gray seeds, *E. glyptlosperma,* is found in similar areas.

208 DESERT TEA or SQUAW TEA. *Ephedra viridis.* There are a number of kinds of desert teas but all have tannin-filled green or gray-green stems and small triangular scale-like leaves. Some plants are male, others female. The blossoming male cones, with their yellow protruding stamens, often occur in huge numbers and make the plants very showy. Tea is made by boiling either the green or dried stems.

209 DEVIL'S CLAW. *Martynia proboscidia.* Large-leafed annual herb with showy yellow flowers and large pods, each ending in a curved hooked beak. This beak splits into two hooked spines which sometimes become entangled in the wool of grazing sheep. There are quite a number of related species both in the United States and in Mexico.

210 NEVADA SQUAW TEA. *Ephedra nevadensis.* Recognized from other "desert teas" by its blue-green or gray-green stems and by the occurrence of its scale-leaves in twos. Southeastern California and southeastern Nevada to Arizona, New Mexico, western Texas, and northern Mexico.

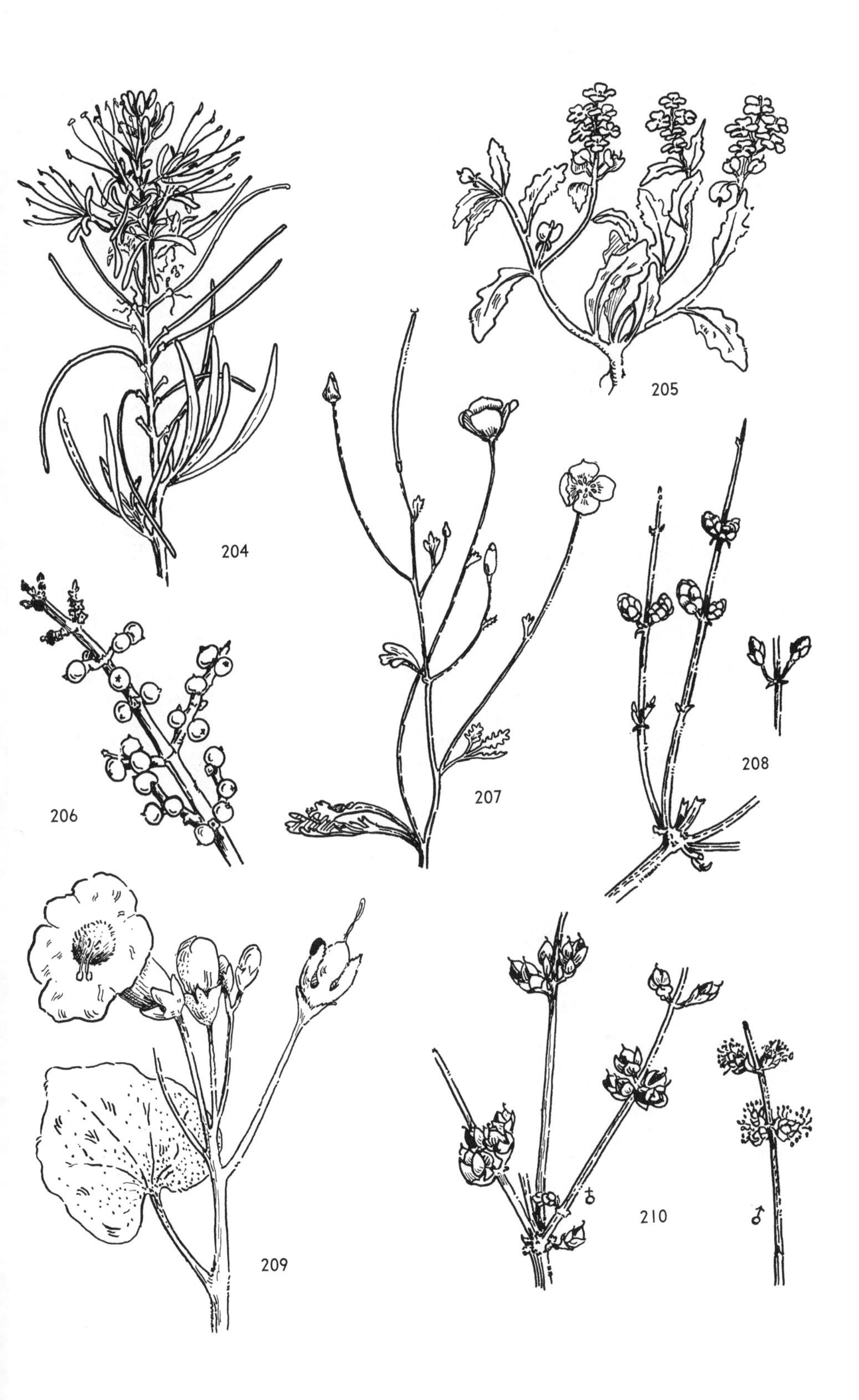

204
205
206
207
208
209
210
♀
♂

211 BLAZING STAR. *Mentzelia nitens.* Annual of sandy areas. The whitish stems have scaling bark; the flowers are yellow. February to May. Southern Utah, Arizona, and southeastern California.

212 GULF MENTZELIA. *Mentzelia adhaerens.* Low rounded bush, often covered with handsome light-yellow flowers and prickly leaves and fruits. Along the Gulf of California both in Sonora and Baja California from San Felipe southward.

213 MENTZELIA. *Mentzelia affinis.* Conspicuously whitish-stemmed, yellow-flowered annual from southern Arizona and southern California deserts. Open places, generally in gravelly soils.

214 BLADDER POD. *Isomeris arborea.* Common much-branched shrub with yellow wood, ill-scented leaves, handsome yellow flowers, and inflated green fruits. Found blooming throughout much of the year when water is available. A favorite source of nectar for bees and hummingbirds. Frequent in sand washes and about springs of the Colorado and Mohave deserts and arid areas of Baja California and western Sonora.

215 DESERT ROCK PEA. *Lotus rigidus.* Low "bush" with several erect stems 1 to 2 feet high. A showy species because of bright canary-yellow flowers. Rocky canyons and hillsides from Death Valley to the Colorado Desert, east to Utah and Nevada.

216 SPIDER FLOWER. *Cleome lutea.* An erect annual with yellow flowers. Southern Arizona to eastern California, mostly in high deserts.

217 BROAD-BANNERED DALEA. *Dalea megacarpa.* Shrub, less than 3 feet high, with both leaves and branches covered with whitish hairs. Flowers with yellow petals, later turning brown. Flower banner three to four times as broad as other petals. Many large pustules on fruit and branches. In sandy washes of Sonora and northeastern Baja California.

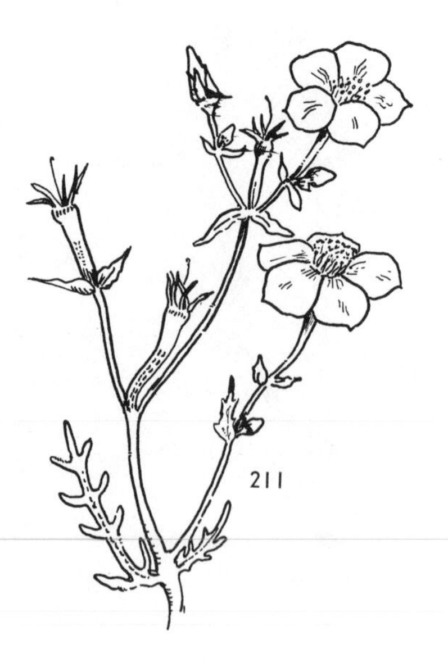

211

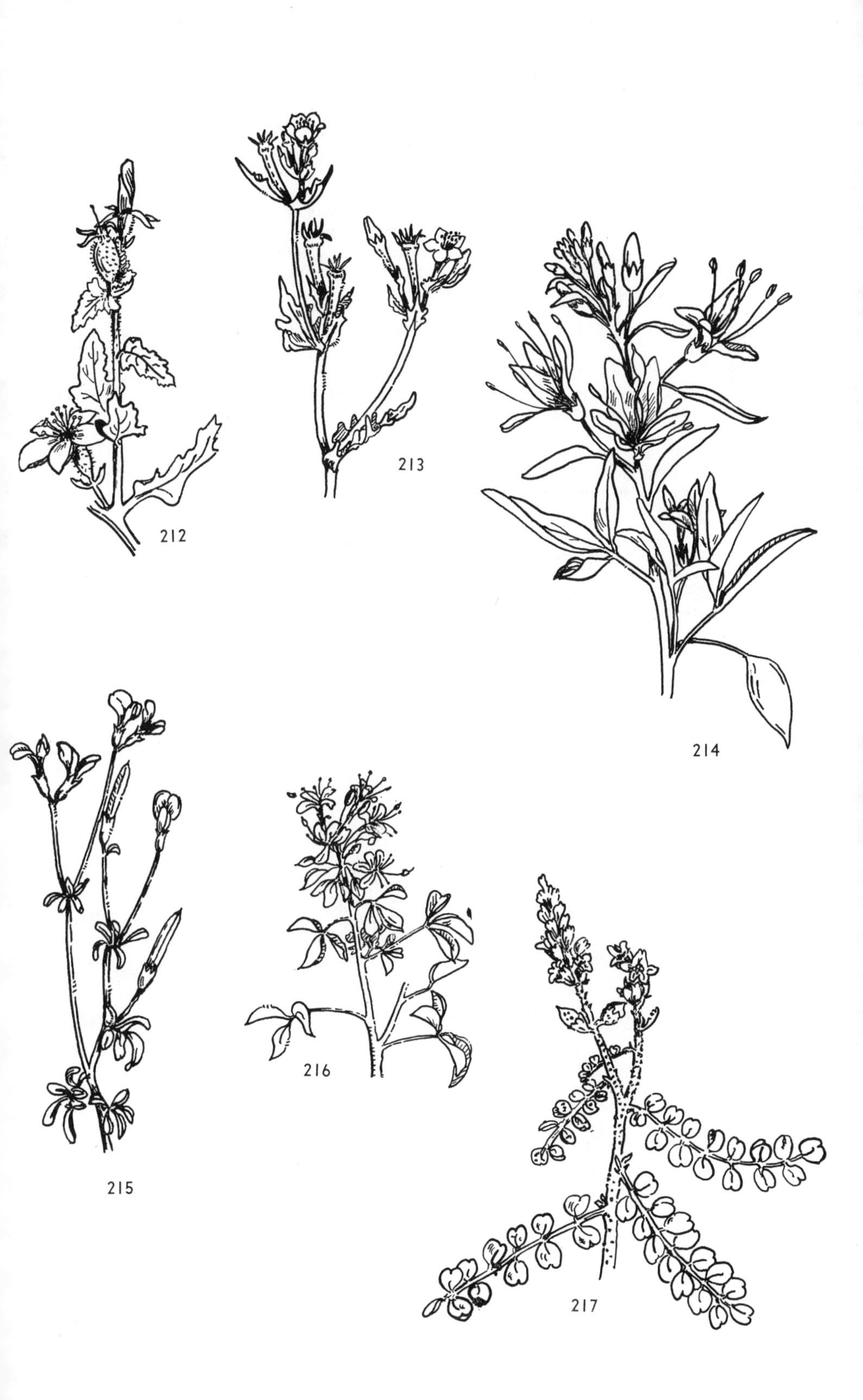

212
213
214
215
216
217

218 MESQUITE. *Prosopis juliflora.* In size, from shrubs to trees 30 feet high (in Mexico). Southern Kansas to southeastern California and Mexico (also, southern South America).

219 VELVET-LEAVED MESQUITE. *Prosopis juliflora velutina.* This southern Arizonan and Sonoran mesquite with neatly regular leaves has seed pods and young stems covered with minute, soft, downy hairs. It is often a puzzler to one who is accustomed to the smooth-leafed variety. The pods have very short beaks.

220 PALO BLANCO. *Acacia willardiana.* A very slender tree, sparsely branched, with white to yellow peeling bark and few "leaves." What appear like leaves are but broadened leaf-stems or cladophylls. Found on rocky ridges, especially on north slopes, of southern Sonora and Baja California. Flowers in pale-yellow spikes.

221 CAT'S-CLAW. *Acacia greggii.* A shrub distinguished by its thin-walled compressed pods and armor of stout, curved, claw-like spines. Aptly called "wait-a-minute bush" or "tear-blanket." The flowers are a valuable source of honey. West Texas to Nevada, southeastern California, and northern Mexico.

222 BIG-THORN ACACIA or CUCHARITAS. *Acacia cochliacantha.* Shrub or small tree known for its large boat-shaped thorns. The yellow flowers occur in globose heads. The long pods are blackish brown when ripe. One of the dominant shrubs of the southern Sonoran plains. Chihuahua, Sonora, and Baja California in foothills.

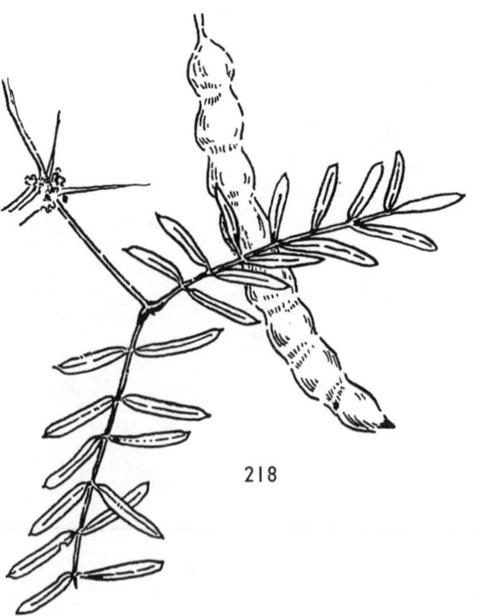

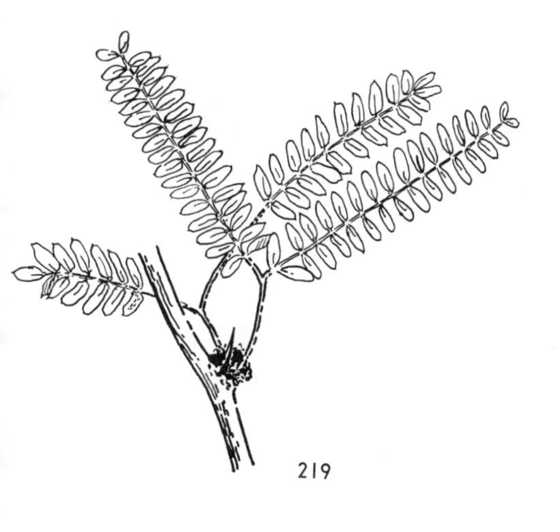
219

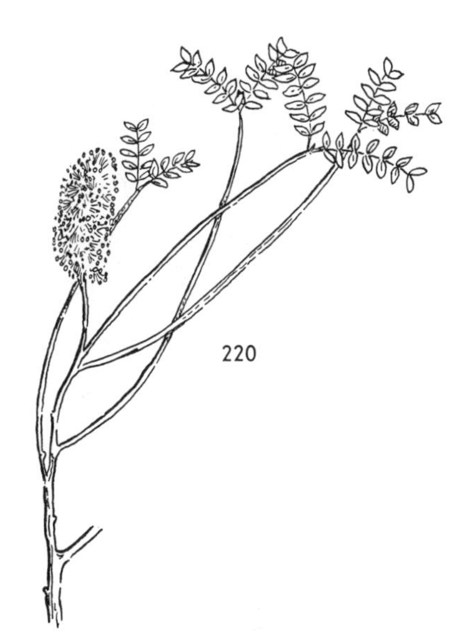
220

221

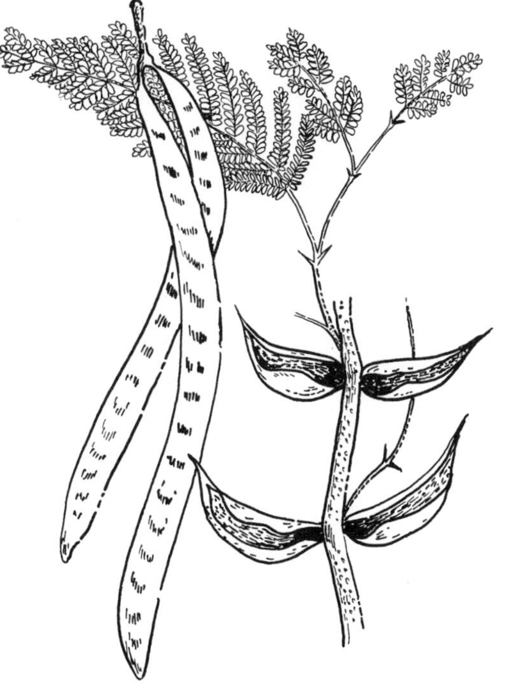
222

223 LITTLE-LEAF PALO VERDE. *Cercidium microphyllum.* The common palo verde of Arizona and Sonora and parts of northern Baja California. A most handsome tree when in spring it is covered deep with myriads of yellow flowers. Its branches are not so brittle as those of the palo verde found farther west; hence there is little litter on the ground beneath it. The leafy branchlets often end in spines. Frequent on rocky hills and arroyo banks.

224 TORNILLO or SCREW BEAN. *Prosopis pubescens.* One of the common shrubs or small trees of the alluvial soil of river bottoms and lagunas. The wood is hard and used extensively for fence posts. The strange twisted pods are very sweet when chewed. Western Texas to Arizona and California; northern Mexico, including Baja California.

225 PIOHITO or PALMER'S POINCIANA. *Caesalpinia palmeri.* Shrub or small tree with handsome yellow flowers in short racemes. The bean is greenish-yellow tinged with purple. Baja California, Sonora, Sinaloa.

226 WHITE THORN. *Acacia constricta.* Spreading hemispherical shrub with golden-yellow flower heads. The prominent thorns are conspicuous because white. The flattened pods are light reddish brown. Grasslands of Arizona to western and southern Texas; Mexico.

227 BLUE PALO VERDE. *Cercidium floridum.* A small to large, broad-crowned tree of the sandy washes of the Sonoran Desert of Baja California and southeastern California. The bright green of its trunk and branches forms a beautiful contrast to its adorning yellow flowers of the spring season. Strong winds frequently break off the brittle branches. The generic name comes from the Greek word for a shuttle, given in reference to the spindle-shaped seed pods.

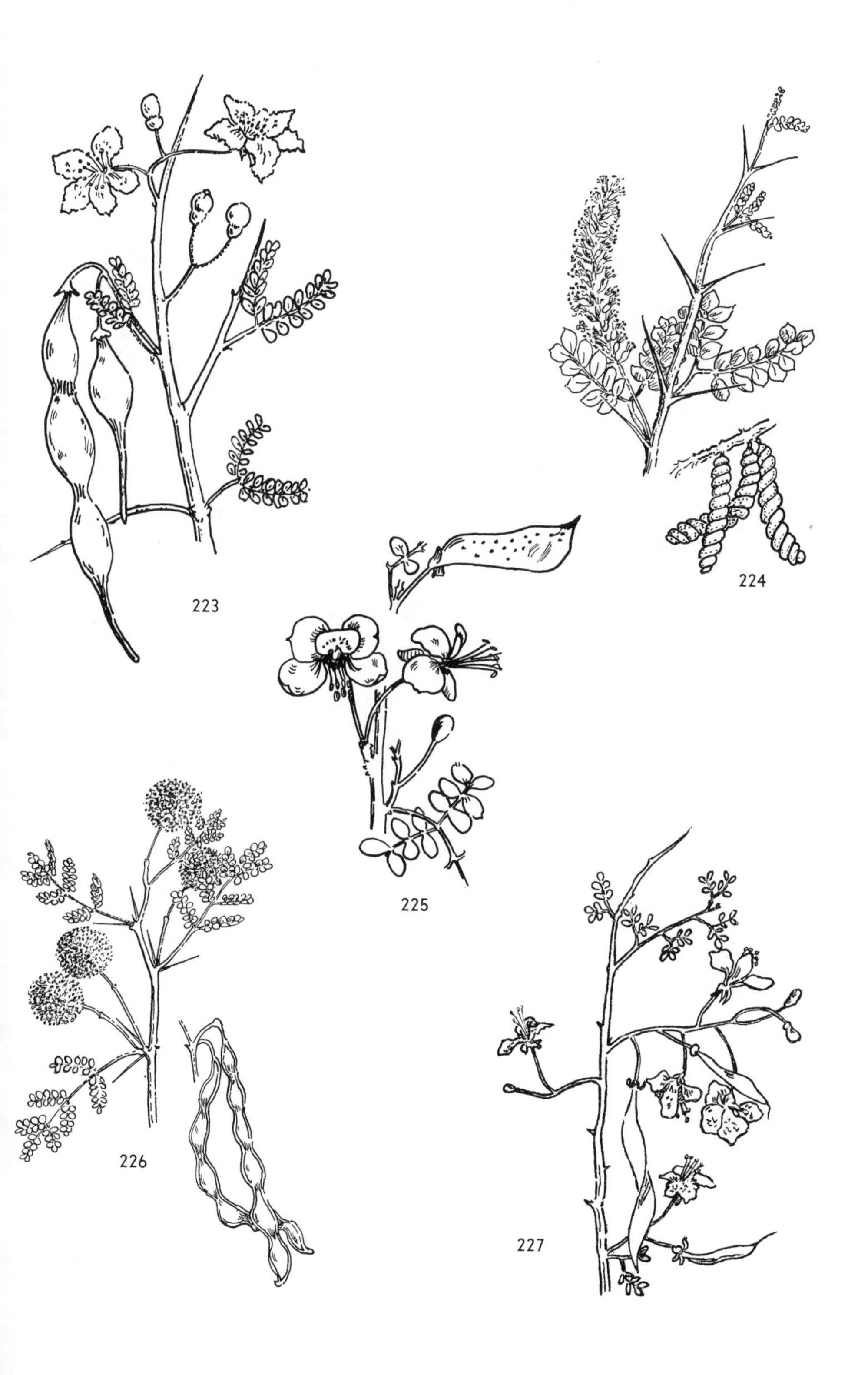
223
224
225
226
227

228 GALLINITA. *Mascagnia macroptera.* Climbing or erect shrub with showy yellow "creped" flowers and peculiar winged seeds. Baja California and Sonora, southward to Sinaloa.

229 COUES' CASSIA. *Cassia covesii.* The somewhat woody plants are densely hairy. The yellow flowers are borne in the axils of the leaves. Low altitude deserts of New Mexico, Nevada, Arizona, California, and northwestern Mexico.

230 BUSH PENSTEMON. *Penstemon microphyllus.* A handsome yellow-flowered shrub, 3 to 6 feet high, found in rocky places of the desert mountains of the Colorado, southern Mohave deserts; eastward in Arizona. Flowers in March.

231 DESERT GROUND-CHERRY. *Physalis crassifolia.* Flower pale tawny-yellow. Often seen in narrow canyon bottoms and washes. The plants have long radish-like, water-storing roots. Widespread from southeastern California to Arizona and northern Mexico.

232 JANUSIA. *Janusia californica.* Slender, climbing shrub, with ovate or oval leaves and small yellow flowers with five clawed petals. The fruit consists of two or three red-tinged, winged parts called samaras. Baja California, Sonora, and Sinaloa. *Janusia gracilis*, with narrow linear leaves, is found not only in northern Mexico but also in western Texas and westward to southern Arizona.

233 BLACKBRUSH. *Coleogyne ramosissima.* Low, tough, twiggy shrub of higher slopes (3,000–4,500 feet) and valleys of the Mohave, Colorado, and Great Basin deserts; northern Arizona. What appear to be yellow petals are sepals. Both sheep and goats are fond of it.

234 DESERT CASSIA. *Cassia armata.* In winter, when leafless, a most woebegone shrub, 1 to 2 feet high, but in spring, after rains, one of the desert's most attractive plants. The numerous bright-yellow flowers are borne in loose panicles on the smooth gray-green stems. Deserts of western Arizona, southern Nevada, and southeastern California.

228

229

230

231

232

233

234

235 DESERT DANDELION. *Malacothrix californica.* Abundant, freely flowering, short-statured annual with canary-yellow flowers, often with a red "button" in the center. Common in open sandy deserts of the Colorado and Mohave deserts.

236 PECTIS. *Pectis papposa.* Low, pungent-odored summer annual with yellow flowers. Its seeds germinate only at above average temperature; therefore it seldom if ever is found growing in early spring. The plant was used by some of the Indian tribes as a flavoring. Sometimes it is eaten, when dried, by horses.

237 CALTROP or ARIZONA POPPY. *Kallstroemia grandiflora.* A pretty herb with large, many-stamened, yellow to deep-orange flowers, each with ten- to twelve-celled ovary. Found from Texas to Arizona and Mexico. Caltrop is a summer annual. It is closely related to the creosote bush.

238 CREOSOTE BUSH. *Larrea divaricata.* One of the commonest shrubs of the Sonoran Desert. Its odor, distinctly pleasant to desert people, is especially noticeable after rains. Often found in almost pure stands from southwest Utah and Nevada to Arizona, Texas, and northern Mexico.

239 WOOLLY DAISY. *Eriophyllum wallacei.* Common annual often found in dense stands on sandy soils of the deserts of southern Utah, southern Nevada to western Arizona, and southern California.

240 SONORAN DEER APPLE or IBERVILLEA. *Maxicmowiczia sonorae.* A great water-storer, among the oddest of desert plants. From the turnip-like root, often half exposed, rise, in wet years and dry alike, the long slender, climbing stems which bear the dissected leaves and small flowers. The sexes of the flowers are separate; fruit is a round red berry. Plants generally found under trees, among whose branches the vines climb. The pulp of the root has an "excruciatively bitter taste." In the variety *peninsularis* of Baja California the upper exposed root often has one to several bottleneck upward extensions which bend over at the top.

241 PAPACHE. *Randia thurberi.* Shrub of the madder family, 5 to 9 feet high, with tubular flowers and small, greenish-yellow fruits. These when ripe are filled with numerous seeds embedded in a sweet, deep-bluish or purple pulp. The birds are so fond of them that one seldom finds a fruit that has not had the interior eaten out.

242 VISCAINOA. *Viscainoa geniculata.* Shrub, 3 to 10 feet high, with crooked branches and ash-green leaves, usually entire. The flowers are large, yellowish white; the fruit is a four-lobed capsule. *Viscainoa pinnata* has 3- to 5-foliate leaves and bright yellow flowers. Central Baja California.

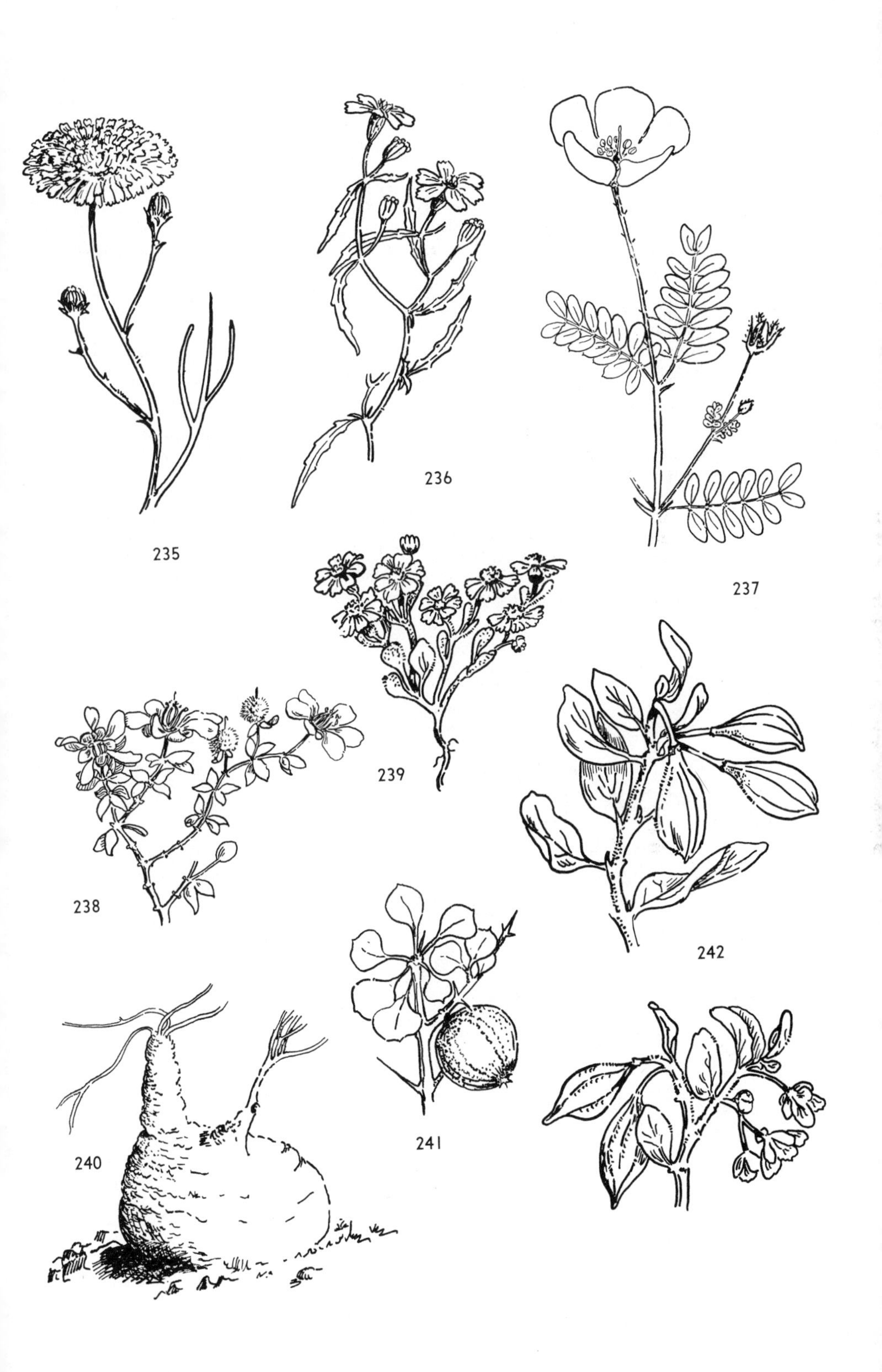

235
236
237
238
239
240
241
242

243 YELLOW CUPS. *Oenothera brevipes.* One of the handsomest of the yellow evening primroses. A free-flowering annual with a number of reddish stems; leaves principally basal. Found in open deserts and washes, southern Nevada, western Arizona, and southeastern California.

244 MILKWEED. *Asclepias subulata.* A rush-like, naked-stemmed plant; stems usually many (20 to 300 or more) and less than 3 feet high. Flowers pale green. Rocky washes; also dry slopes. Southern Arizona, southeast California, and northwest Mexico. The dried milk may contain up to six per cent of rubber.

245 INFLATED-STEM BUCKWHEAT or DESERT TRUMPET. *Eriogonum inflatum.* A plant 1½ to 4 feet high, with peculiar inflations of the upper, hollow flowering stems. The young stems taste like sheep sorrel. Utah to southern California, Arizona, and Baja California.

246 TECOMA. *Tecoma stans.* Opposite-leaved ornamental shrub with handsome bright-yellow flowers. The seeds are flat with thin paper-like wings. Southern Arizona, southern New Mexico, and southward into Chihuahua (thence to tropical America).

247 YELLOW FELT-PLANT. *Horsfordia newberryi.* Tall, slender, somewhat woody plants with a dense felty covering of yellowish hairs. In some species the flowers are yellow, in others pink. In the pink-flowered felt-plants the stems may reach a height of almost 10 feet. Southeastern Arizona, northern Sonora, Baja California, and southern California in rocky soils.

243

244

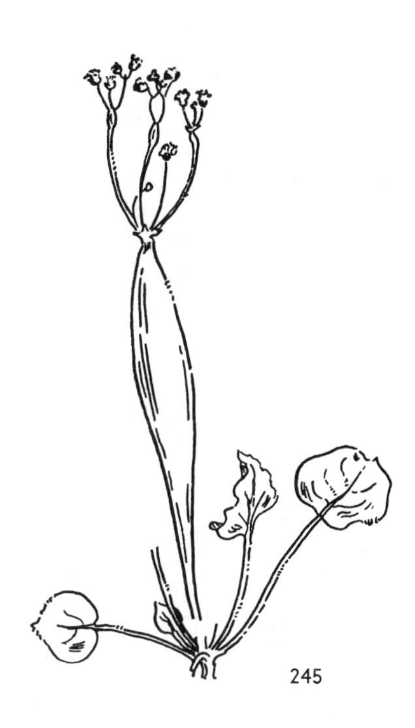
245

246

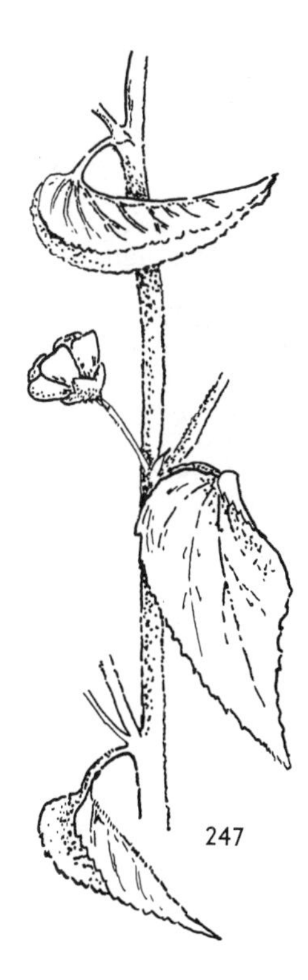
247

248 GOLDEN GILIA. *Gilia aurea.* Among the handsomest of the small gilias and often found in large aggregations, making a carpet of gold on the desert floor.

249 FREMONT BURRO BUSH. *Franseria deltoidea.* Plants are woody at base, leaves whitish beneath and with distinct stem. The numerous burrs bear up to 20 strongly flattened spines. Abundant on plains of south-central and western Arizona to Sonora. First collected by John C. Frémont along the Gila River. Associated with creosote bush.

250 GHOST FLOWER. *Mohavea confertiflora.* A handsome, annual, desert figwort with pale-yellow, purple-dotted flowers. Locally abundant in sandy flats and stony slopes of detrital fans in western Nevada, western Arizona, southeastern California, and northern Baja California on the gulf side.

251 BURRO BUSH or WHITE BURR SAGE. *Franseria dumosa.* Next to the creosote bush, the Sonoran Desert's most widespread shrub. The leaves are silvery green, the flowers inconspicuous. It is almost as drought resistant as the creosote bush and often is found growing in the open spaces between. Southern Utah to southeastern California and northwestern Mexico. Both horses and burros are very fond of it.

252 CAMPHOR-ODORED BURRO BUSH. *Franseria camphorata.* In among the large tree cacti and cirios of northern Baja California grows this attractive bush with camphory, resin-scented, finely cut, silvery green leaves and small burr-like fruits. Flowers inconspicuous. First discovered on Guadalupe Island.

253 HOLLY-LEAF BURRO BUSH. *Franseria ilicifolia.* Handsome large-leaved, rounded sub-shrub of low rocky canyons and sand washes of the hot Yuman and Colorado deserts (Gila Mountains in Arizona, Chuckawalla Mountains in California, desert slopes of Sierra Juárez and Sierra San Pedro Mártir in Baja California).

254 THREE-LEAF BARBERRY or ARGILLO. *Berberis trifoliata.* Spiny-leafed, tall shrub with inner bark yellow; roots also yellow. The leaves are thick, rigid, and pale green, especially beneath. The round, red to bluish-black fruit has several seeds and is edible. The wood is sometimes used for tanning. Western Texas to New Mexico in low dry hills; also on the Chihuahuan Desert south to Durango.

255 CANDELILLA. *Euphorbia antisyphillitica.* Also called waxplant because it yields a wax used in candle making. The numerous pale-green, milky-juiced, wax-covered stems, each from 1 to 3 feet high, are almost leafless and occur grouped in characteristic close clusters. To obtain the wax the stems are boiled in water, and the wax is allowed to rise to the surface; refining makes it white. Western Texas and northern and eastern Chihuahuan Desert of Mexico.

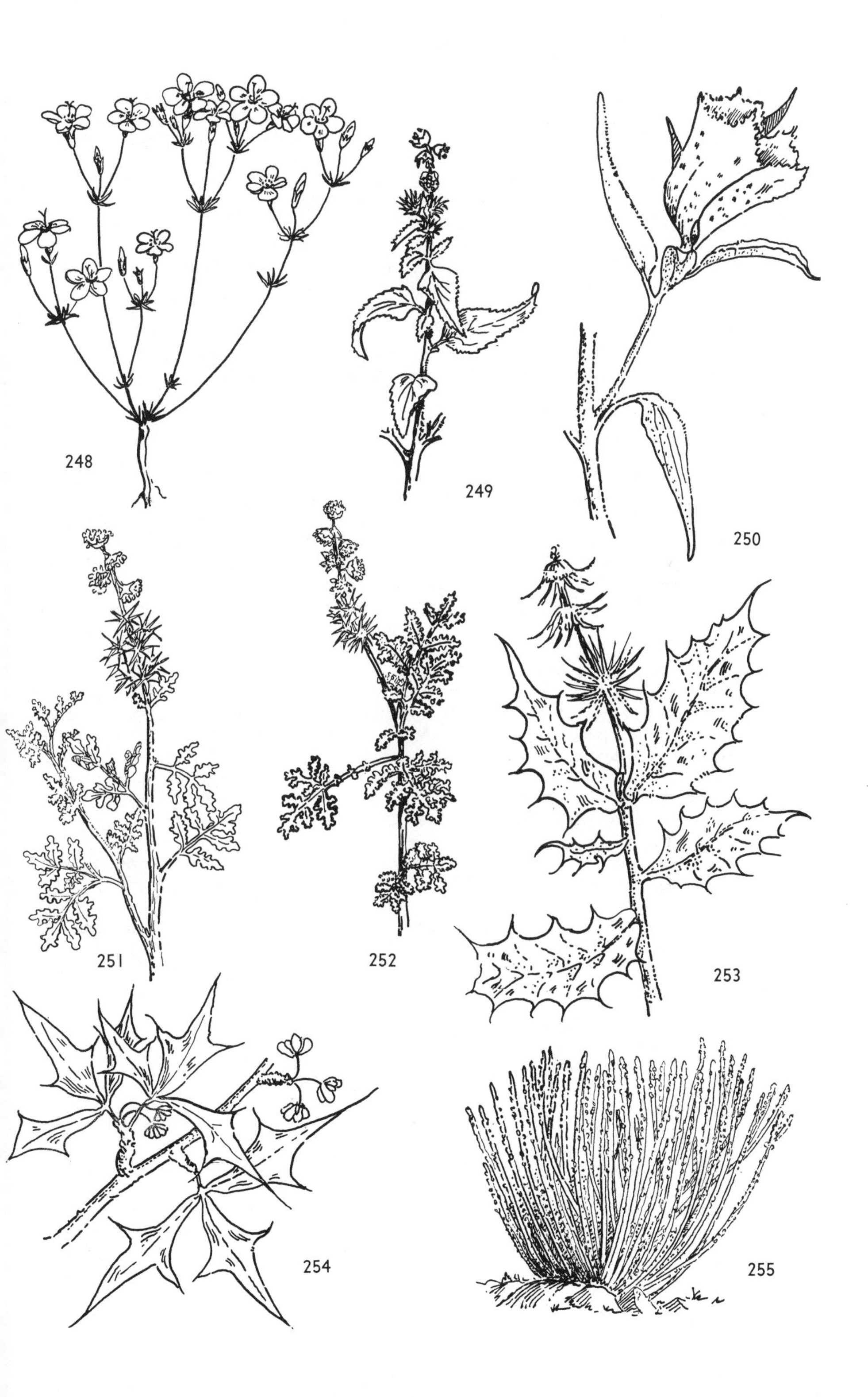
248
249
250
251
252
253
254
255

256 SCALE - BROOM. *Lepidospartum squamatum.* Rigid, green, broomlike shrub with small appressed scale-like leaves. Common in stony washes from California and Nevada to Arizona. A water-indicator. Blooms throughout summer.

257 GOAT NUT or JOJOBA. *Simmondsia californica.* A 2- to 4-foot shrub whose branches are almost hidden by the ovoid, leathery, gray-green leaves. The small nuts taste much like filberts. The desert rodents harvest most of them. By weight the nut is almost 50 per cent commercially valuable liquid wax. Peculiarly, this wax is wholly undigestible, so that there is relatively little nourishment gained by eating the nuts.

258 FINGER-LEAVED GOURD. *Curcubita palmata.* This beautiful gourd is best known from the sandy plains of the low deserts of Arizona, New Mexico, and adjoining Chihuahua. Sometimes called coyote melon.

259 SINGLE-LEAF NUT PINE. *Pinus monophylla.* This piñon or nut pine is common over the low arid mountains of the Mohave and southern Great Basin deserts. In parts of Arizona and New Mexico its place is taken by the TWO-NEEDLE NUT PINE (*Pinus edulis*). In similar situations in the Chihuahuan Desert grows the THREE-NEEDLE NUT PINE (*Pinus cembroides*), while in the mountains of northern Baja California grows the beautiful FOUR-NEEDLE PIÑON (*Pinus parryi*). In all of these arid-mountain pines the young trees have conic crowns but the older specimens are characterized by rounded to flat-topped brushlike tops, and numbers of more or less horizontal main side branches. The large, well-flavored nuts are sought by birds, rodents, and man.

260 BUFFALO GOURD or CALABAZILLA. *Cucurbita foetidissima.* Conspicuous trailing perennial herb with thick water-storing root, large broad triangular-ovate leaves, yellow flowers, bright yellow gourds. Nebraska to Texas, Arizona, southern California, and Mexico. The two other nearly related gourd species are easily identified by the leaves. *Cucurbita digitata* occurs from southwestern New Mexico and northern Mexico to southeastern California. *Cucurbita palmata* is known from the deserts of southwestern Arizona, southeastern California, and northern Baja California.

261 HUERO DE GATO. *Cardiospermum halicacabum.* Vine with peculiar inflated, papery, three-angled pods. Each division contains a brownish-black, rounded seed with a yellowish-white "eye" or spot. The pods vary much in size from place to place. The yellowish or white flowers are irregular in form and have four petals. Coiling tendrils aid the plant in maintaining a hold as it climbs. Northern Mexican mainland and Baja California.

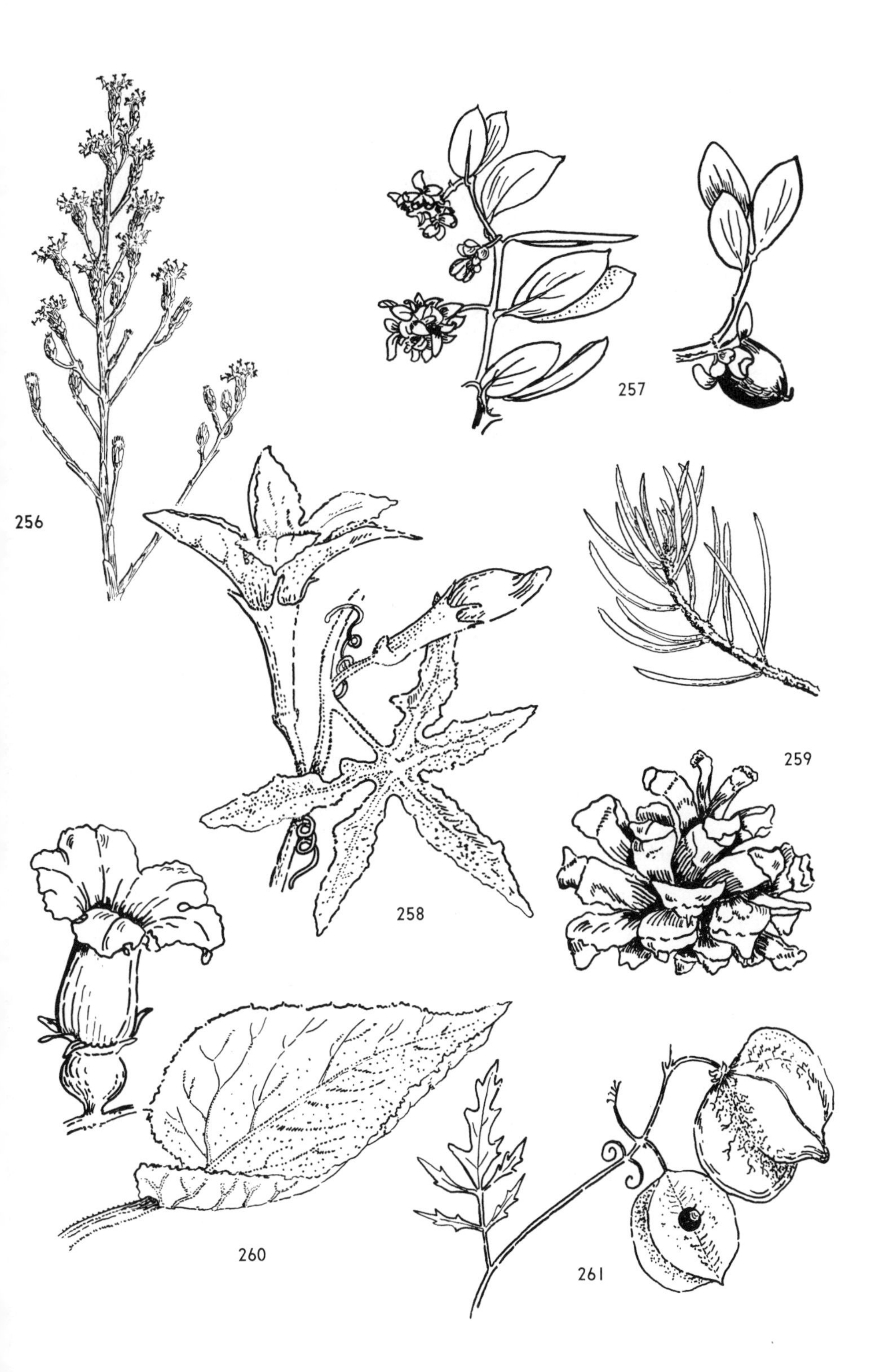
256
257
258
259
260
261

262 LINEAR-LEAVED GOLDENBUSH. *Haplopappus linearifolius interior.* A rather common bush of rocky hillsides and gravelly slopes; yellow flowers conspicuous in spring. At other times it has little of beauty to boast. Widespread from Colorado to California and Baja California.

263 TRIXIS or PICHAGA. *Trixis californica.* Low green shrub with golden-yellow flowers; most commonly found in rocky washes and on stony slopes. The herbage is exceedingly bitter; the leaves are dotted with glands on the lower surface. This is but one of the numerous species of trixis found in Mexico. Sonora to Coahuila, Zacatecas; Texas to southeastern California.

264 GROUNDSEL. *Senecio douglasii.* Bushy shrub with leafy branches and yellow flowers. Often seen in desert washes and alluvial plains from southern California to Baja California and Sonora. In the Death Valley region look for the handsome *Senecio monoensis* which grows as a more compact bush.

265 HOJASE, TAR BUSH. *Flourensia cernua.* Resinous, bitter-tasting shrub from 3 to 6 feet high with nodding yellow heads of flowers. The crushed herbage has a hop-like odor. The Mexicans use the dried leaves and flower heads to make a remedy for indigestion. Very common in the northern Chihuahuan Desert where limestone soils are found. Often associated with guayule. Found also in western Texas to southeastern Arizona.

266 COOPER GOLDENBUSH. *Haplopappus cooperi.* Low, flat-topped shrub often occurring in almost pure stands on the flats of the Mohave Desert. The flowers are bright yellow.

267 BRITTLE BUSH or INCIENSO. *Encelia farinosa.* Widely distributed hemispherical, resinous-stemmed shrub with gray-green leaves. In Baja California is a form with purple flower-disks. This drought-resistant plant is widespread from southern Nevada and southeastern California to Baja California, Sonora, and Sinaloa.

268 DESERT FIR or PIGMY CEDAR. *Peucephyllum shottii.* A rounded, dark-green, evergreen shrub of medium height with abundant fir-like leaves and light-yellow, bell-shaped flower heads. Rocks, banks, hillsides, and washes, southern Nevada to Mohave and Colorado deserts, extreme western Arizona, Baja California, and Sonora.

DESERT PLANTS (*yellow*)

263
264
265
266
267
268

269 VIGUERA. *Viguera deltoidea.* Shrub with much the appearance of brittle bush except that its stems are more slender and its leaves green and more or less triangular. Usually found in gullies and rocky areas. Widespread from Sonora and Baja California to Arizona, southern Utah, Nevada, and southeastern California.

270 DUNE SUNFLOWER. *Helianthus niveus.* A small-statured sunflower with herbage silvery to white because of the many small hairs covering it. Rare to abundant in sandy areas after plentiful rains of winter. Colorado Desert of California and Baja California.

271 GUAYULE. *Parthenium argentatum.* A low-branching shrub with narrow, lanceolate, few-lobed leaves, silvery on both sides. The flower heads are small, each with five short ray flowers. Only the ray flowers are fertile. This is the famous rubber-plant of the Chihuahuan Desert; the stems yield a high percentage of rubber. Large fields of guayule were cultivated in the western United States as a source of rubber during World War II.

272 GOLDENHEAD. *Acamptopappus sphaerocephalus.* A small but often common shrub of flats and benches of the western Colorado and Mohave deserts. The rounded flower heads, light-yellow to golden-yellow in color, often cover the entire bush. Search among its branches for the pure-white, inch-long inflated cases of the goldenhead bagworm.

273 NEVADA VIGUERA. *Viguera multiflora nevadensis.* Yellow-flowered perennial with brownish stems 1–1½ feet high. Common on deserts of southern Nevada and in eastern California.

274 HOFFMANSEGGIA. *Hoffmanseggia microphylla.* Small shrub with wandlike branches. The petals are yellow or yellow and red. The fruit is a small, flat, broad pod covered with stalked glands. Areas of sand or clay in the lower Colorado River drainage to Baja California and Sonora.

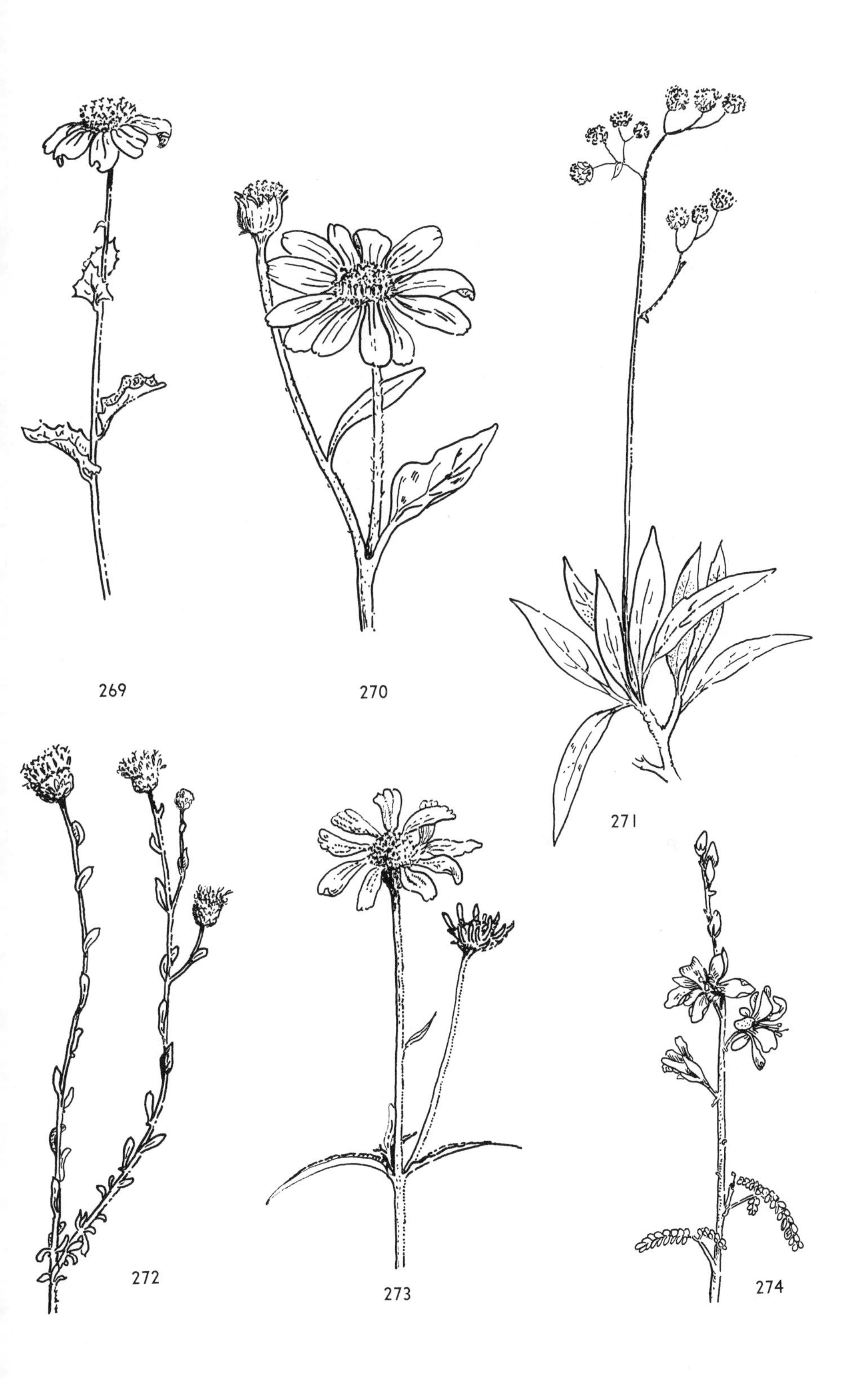
269
270
271
272
273
274

275 SCALE BUD. *Anisocoma acaulis.* Pale-yellow, freely flowering annual of sandy areas. A brown stripe runs down the center of each bud scale. California desert to Nevada and Arizona.

276 GOLD DOLLARS, PAPER DAISY, or DESERT MARIGOLD. *Baileya multiradiata.* Low woolly annual herb with yellow flowers set on long peduncles. The floral rays are papery and turn down with age. A common and very showy roadside plant; also on sandy plains. Found from deserts of western Texas to northern Mexico and Utah and California.

277 BIGELOW'S COREOPSIS. *Coreopsis bigelovii.* A very profuse-blooming common annual of open sandy flats of the southern California deserts. The leaves are all basal. Rather similar to *Coreopsis douglasii* of western Arizona and the Colorado Desert of California and Baja California.

278 DESERT SUNFLOWER. *Geraea canescens.* A showy annual often occuring in dense stands on desert flats. The large yellow flowers are sweet-odored. Desert areas of southern Utah to southern California, southern Arizona, and Sonora.

279 LAX FLOWER. *Baileya pauciradiata.* Erect herb, ¾–1½ feet high, covered with loose hairs. The lemon-yellow ray flowers turn down with age, giving a lax appearance to the flower heads. Mohave, Colorado, and Yuman deserts.

280 SNAKE'S-HEAD. *Malocothrix coulteri.* Flowers light yellow, subtended by papery margined phyllaries, striped with brown. A handsome annual of sandy areas. Appears in March and April, southwestern Utah to southern Arizona and southern California (also Argentina—How did it get there?).

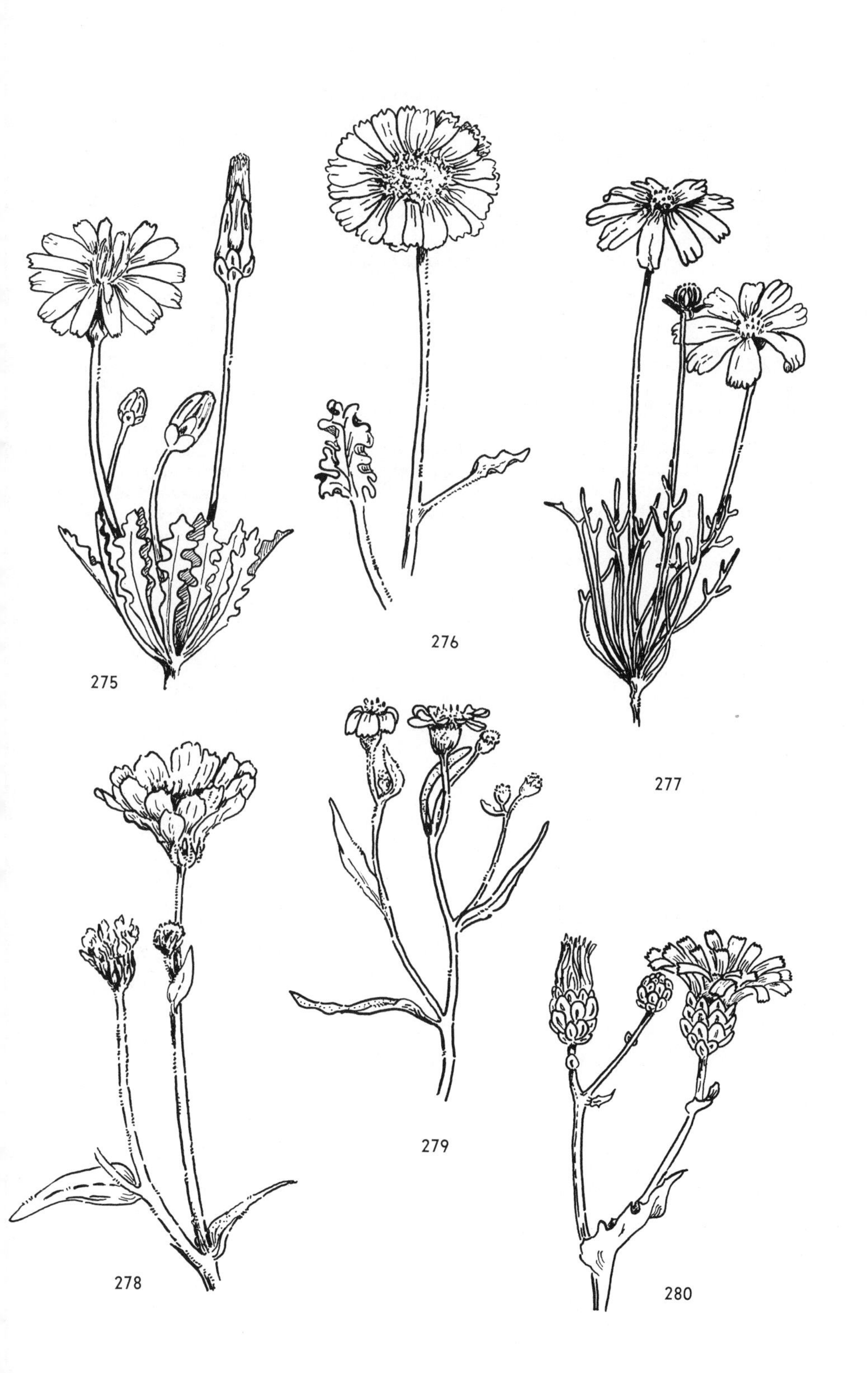
275
276
277
278
279
280

281 YELLOW FIDDLENECK. *Amsinkia intermedia.* Bristly, erect plant with yellow-orange flowers. Found abundantly on dry sandy or gravelly soils, particularly about the bases of creosote and cat's-claw bushes. Mohave, Colorado, and Yuman deserts.

282 DESERT MALLOW. *Sphaeralcea ambigua.* One of the handsomest of low desert shrubs and frequently met with. The apricot-colored flowers are often produced in marked numbers. Common to both California deserts, to Arizona, Nevada, and northern Baja California.

283 APRICOT MALLOW. *Sphaeralcea ambigua subsp.* A smaller-leaved form of the desert mallow, found at higher elevations, generally in rocky areas.

284 KENNEDY MARIPOSA LILY. *Calochortus kennedyi.* Handsome low-growing, thick-stemmed "lily" with rich, deep-orange flowers. Southern Arizona, southeastern California to Sonora.

285 MEXICAN POPPY. *Eschscholtzia mexicana.* A species closely related to the CALIFORNIA POPPY (*Eschscholtzia californica*) and often intergrading with it in Arizona, and, like that species, often in favorable seasons coloring extensive areas with its showy orange flowers. Early in the season the plants hug closely to the ground but later may develop long, leafy stems.

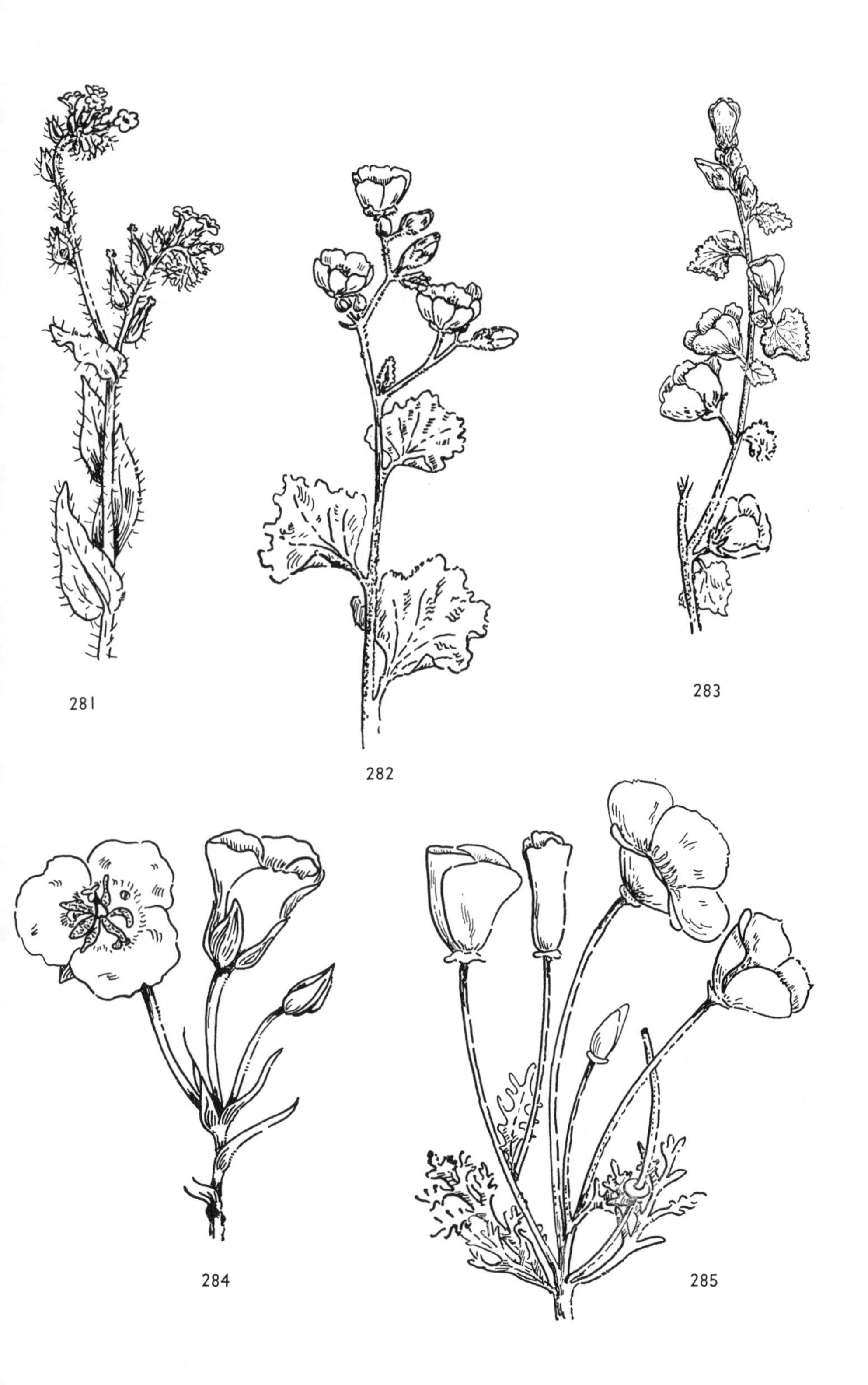
281
282
283
284
285

286 FIVE-SPOT. *Malvastrum rotundifolium.* A handsome annual with globe-like, pinkish-lilac flowers, each of the five petals conspicuously spotted with darker violet-pink. Found in dry sandy places of desert areas of western Arizona, southern California, and southern Nevada.

287 SAND VERBENA. *Abronia villosa.* Trailing-stemmed annual often blanketing wide areas of sandy soil. The numerous rose-purple flowers are very sweet-scented, especially at night. Southern Arizona, southern California, northern Baja California, and Sonora.

288 WINDMILLS. *Allionia incarnata.* Spreading, slender-stemmed annual, with stems and leaves often covered with bits of sand and mica due to the viscid surface. Each apparent flower is really three flowers closely set. Colorado and Mohave deserts.

289 BOERHAAVIA. *Boerhaavia intermedia.* A representative of many species of boerhaavia found on our deserts. Some are annual, some perennial. The flowers are very small. Plains and mesas, generally on sandy or gravelly soils. Strangest of all the species is the perennial NAPKIN-RING BOERHAAVIA of the Death Valley area. The stout stems are adorned with reddish mucilaginous rings.

290 SLENDER-STEMMED BUCKWHEAT. *Eriogonum gracillimum.* Freely branching Mohave Desert annual, up to 10 inches high, with woolly herbage and rose-pink florets. Its flowers are flaring funnel-form in contrast to the more widely spread, quite similar *Eriogonum angulosum* of Arizona and Baja California, which has globular florets.

291 DESERT CALICO. *Gilia mathewsii.* Freely flowering annual, often occurring in dense aggregations or singly in sandy flats. Flowers white to deep pink with patterns of dark specks on the petals of the upper lip. A persistent bloomer late in the season. Colorado and Mohave deserts south to Arizona and Sonora.

286
287
288
289
290
291

292 ELEPHANT TREE or COPAL. *Elaphrium macdougalii.* Deciduous-leafed shrub or small, round-crowned tree of sandy and rocky washes and lower rocky mountains of Baja California from near San Felipe on the Gulf of California southward. It is one of the numerous species of copals and a northern representative of a group of arid tropical trees common in western Mexico.

293 ELEPHANT TREE or TOROTE. *Bursera microphylla.* Spicy-odored shrub or small tree of the dry rocky hills and plains of northern Sonora and northern Baja California. The thick stems exude a red sap used in dyeing. In winter and again in summer the leaves may drop off. The numerous fruits are bluish green.

294 ASHY JATROPHA. *Jatropha cinerea.* Shrub or small tree with brown or whitish papery bark, the small flowers pinkish. The branches are rubbery and full of mucilaginous sap. The puckery juice is used as a remedy for warts. Plains and dry hills of Sonora, Baja California, Sinaloa.

295 WAIT-A-MINUTE BUSH. *Mimosa biuncifera.* Spiny, small-leaved, thin-stemmed shrub of the foothills and plains of the Arizona Upland and Chihuahuan deserts. In spite of its small leaves, deer and goats graze it. Distinctive are the paired prickles of the stems, and the marginal hooks along edge of the fruits.

296 SILKY DALEA. *Dalea mollis.* A low-growing, very hairy perennial with many dark glands. Pleasantly aromatic when crushed.

297 DESERT WILLOW. *Chilopsis linearis.* Not a willow but a wild catalpa known for its handsome clusters of pink flowers, long narrow beans, and narrow leaves. An inhabitant of sandy washes carrying runoff cloudburst waters.

298 FAIRY DUSTER. *Calliandra eriophylla.* A low straggling shrub of gravelly slopes and mesas (500 to 5,000 feet) advertising itself in spring by its handsome pink long-stamened flowers. West Texas to southeastern Colorado Desert and Mexico (Sonora to Coahuila and Puebla). In Baja California are several related species.

292

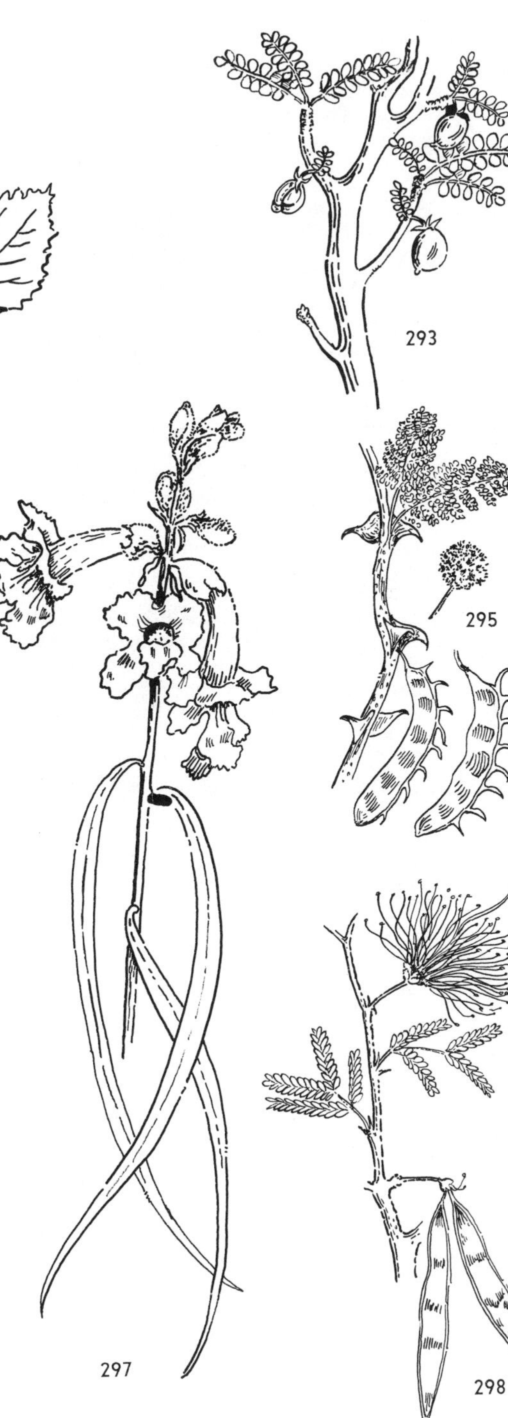

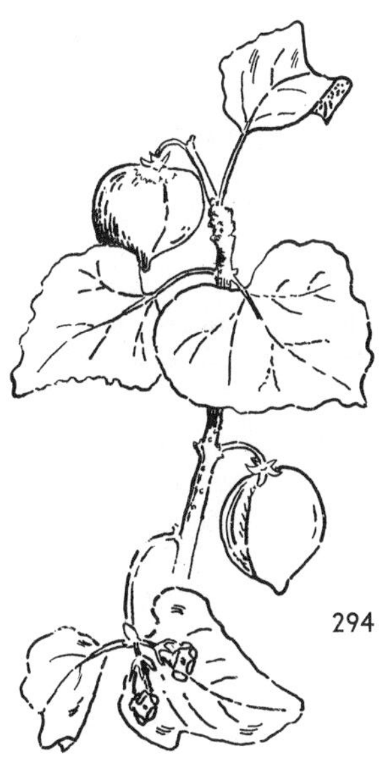

294

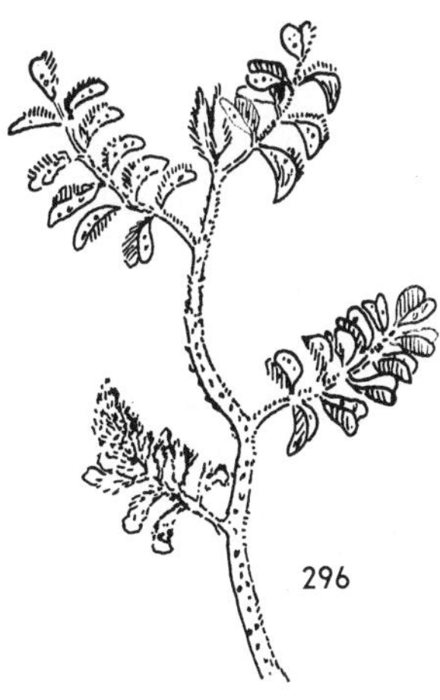

296

299 STRAWBERRY CACTUS. *Echinocereus engelmannii.* A short-stemmed cactus with cylindric ascending joints covered with yellowish to brown spines. The handsome funnel-shaped flowers with ribbed, pointed petals are purple. The ovoid, spiny red fruits are sweet and edible and contain almost-round black seeds. Utah, Arizona, California south to Sonora, and Baja California.

300 CHAIN CACTUS. *Opuntia fulgida.* So named because the persistent fruits hang in chainlike clusters. The pale-green joints are readily detached and sometimes may be seen hanging to the flesh of feeding cattle. Flowers pink. A very common cholla from southern Arizona southward in Mexico to Sinaloa.

301 SENITA (SINITA) or OLD MAN CACTUS. *Lophocereus shottii.* So called because of the head of long, twisted bristles at the ends of the older stems. This species occurs in large clumps of 20 to 100 upright stems. Flowers small, pink, odorless, opening at sunset and withering next morning; fruits globose, usually without spines, edible.

302 PALMER'S BEARD-TONGUE. *Penstemon palmeri.* A tall, few- to many-stemmed plant with large pink, very fragrant flowers. Found in desert mountains and foothills. It does exceptionally well in limestone soils. The delicate odor of the flowers is much like that of apple blossoms. Utah, Arizona, California.

303 FRINGED ONION. *Allium fimbriatum.* The handsome tufts of flowers, pale-rose to purple, are borne on a stem 2 to 3 inches high. There is but a single leaf. Dry slopes and flats of both California deserts. Nearly related onions are found in the Arizona, New Mexico, and Sonora deserts, often covering the ground with their pinkish flowers.

304 LONG-TUBED GILIA. *Gilia brevicula.* Dainty annual with pink, purple-throated flowers. Sands of western Mohave Desert.

299

300

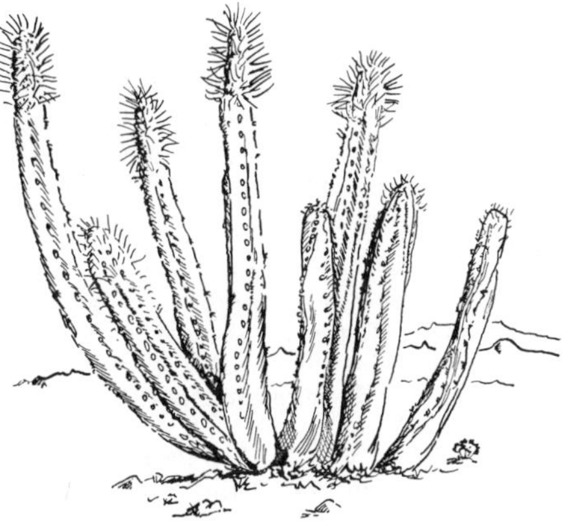

301

302

303

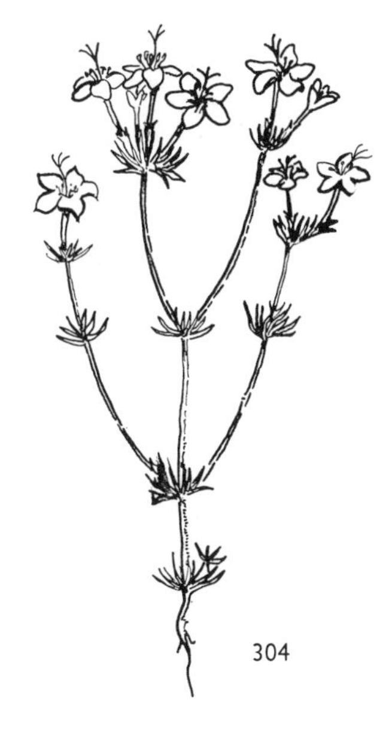

304

305 CHUPAROSA. *Beloperone californica.* A low, rounded shrub with bright nopal-red flowers much visited by hummingbirds. Rarely, one finds a plant with yellow flowers. Chuparosa grows in washes and among rocks of the Colorado Desert southward into Lower California. An early bloomer. The leaves drop during the hot summer.

306 KRAMERIA. *Krameria parvifolia.* Low-branching shrubs, 1 to 2 feet high, which parasitize the roots of near-by plants. These are handsome shrubs in May when the twiggy branchlets are almost covered with numerous fragrant, wine-colored flowers. The heart-shaped fruits are adorned with delicate barbed spines. Colorado and Mohave deserts to Texas and Mexico. An important food of the desert bighorn.

307 BIGELOW MIMULUS. *Mimulus bigelovii.* Flower red-purple. Annual up to 10 inches high. Canyons and washes of both California deserts to Nevada and Arizona.

308 BEAVER-TAIL CACTUS. *Opuntia basilaris.* When in flower, one of the most frequently photographed of all the cacti. The numerous rich magenta flowers grow along the edge of the flat, velvety, gray-green joints. The numerous spines are very short, easily come off when touched, and are hard to extract from one's flesh. After removing the spines the Indians cook the flower buds in their meat stews. This species is widespread from low to high elevations over much of the Sonoran Desert.

309 WHITE RATANY. *Krameria grayi.* Low-branching shrub, parasitic on the roots of other shrubs. Masses of fragrant wine-red flowers adorn the plant in early to late spring. The heart-shaped fruits bear long barb-tipped hairs. Colorado and southern Mohave deserts to Texas and Mexico.

310 TOOTH-LEAF STILLINGIA. *Stillingia paucidentata.* An herb of rank odor with many eight-inch to foot-high stems, bluish-green leaves, and greenish to white spikes of small flowers.

311 TREE OCOTILLO or TOROTE VERDE. *Fouquieria macdougalii.* Tree with thick, short, yellowish-green trunk; crown consisting of brownish branches armed with coarse needle-like spines. The flowers are bright red. The bark is used in washing clothes. Middle and southern Sonora and northern Sinaloa.

312 PAINTBRUSH. *Castilleja, sp.* Most of the paintbrushes are erect perennial herbs which advertise themselves early in the season by their bright-red floral bracts that are much more conspicuous than the flowers accompanying them. Found mostly in desert mountains. Western Texas to southeastern California, southern Nevada, and northern Mexico.

313 SCARLET LOCOWEED. *Astragalus coccineus.* Densely tufted, white, silky leaved "locoweed" arising from a stout taproot. The flowers are bright crimson to scarlet, 1¼–1½ inches long. Desert mountains of Inyo County, California; also mountains of west side of Colorado Desert. All the species of astragalus are known as locoweeds. Only a few are poisonous, causing a "locoed" condition in animals.

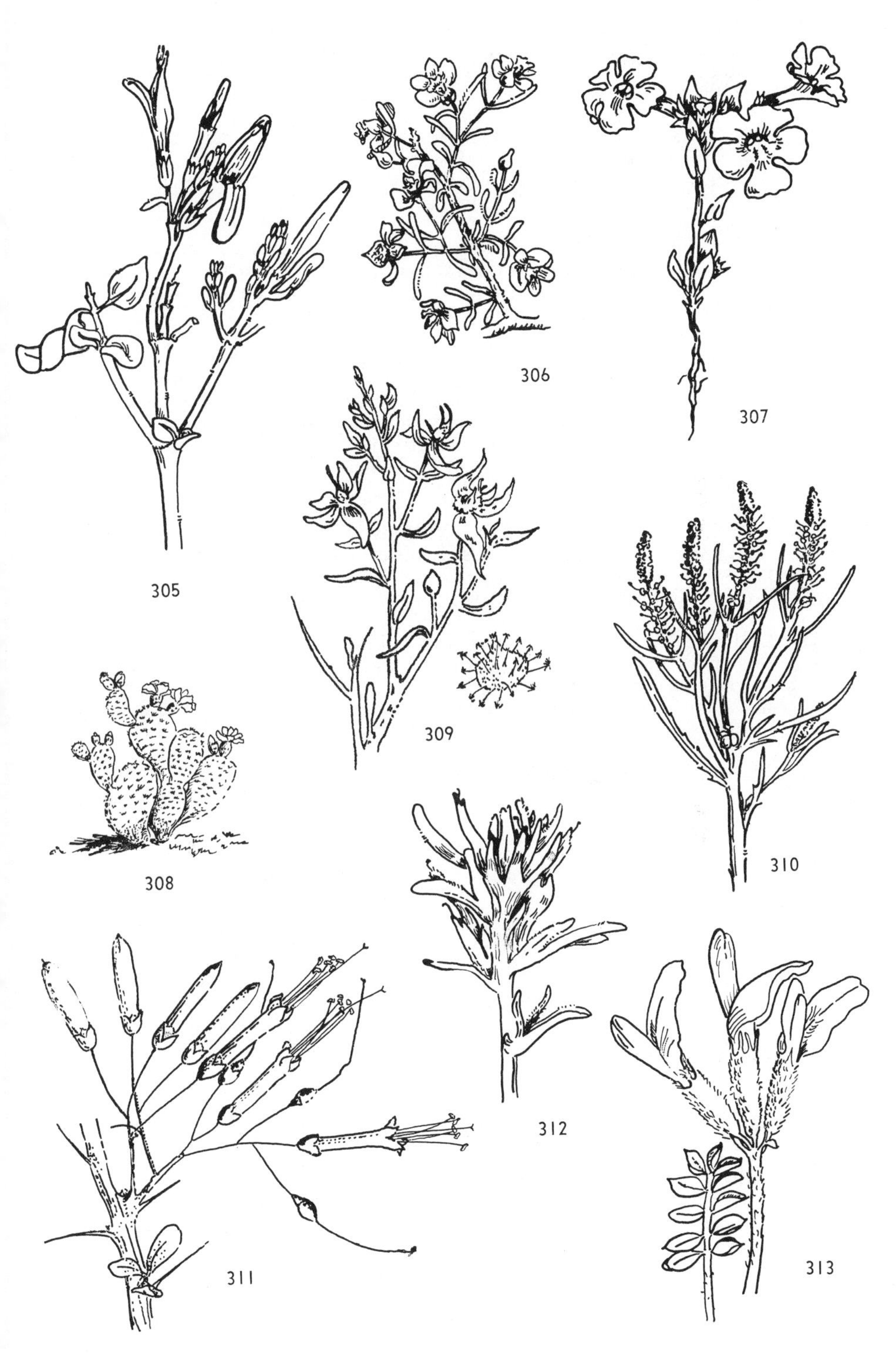

305
306
307
308
309
310
311
312
313

314 MOHAVE ASTER. *Aster abatus.* Handsome perennial with large lilac-colored flowers. It rapidly puts on spiny-toothed, dark-green leaves after winter rains. The lower stems are somewhat woody. Southern Utah, southern Nevada, western Arizona, and southeastern California above 2,000 feet.

315 FAGONIA. *Fagonia californica.* A low half-shrub with purplish-pink flowers and three-parted green leaves. The fruits are deeply cut into five parts. This near relation to the creosote bush inhabits rocky desert hillsides from southern Utah to southern California, Sonora, and Baja California.

316 SQUAW CABBAGE. *Steptanthus inflatus.* So called because the Piute Indians of the Mohave desert used it as a potherb. A particularly conspicuous plant when seen in aggregation. The stems are lemon-yellow to green, the flowers deep purple. An annual.

317 ARROWWEED. *Pluchea sericea.* Slender, willow-like shrub with silvery, silky foliage. Found about seeps and river bottoms where it forms sweet-smelling thickets. The Indians utilized the long straight stems for arrow shafts, also for basket making and for the sides and roofs of shelters. The small flowers are purplish. Nevada to California, Texas, and Mexico. Arrowweed is the shrub forming the shocks of the famous "Corn Patch" of Death Valley and of Saline Valley.

318 EMORY DALEA. *Dalea emoryi.* Densely branched, low rounded shrub, with felted leaves giving off a highly pleasing odor when crushed. The flower heads rubbed between the fingers yield a saffron-yellow dye used by desert Indians. Colorado Desert to Arizona and Baja California.

319 COLDENIA. *Coldenia palmeri.* Coldenias are small, much-branched shrubs, with numerous alternate leaves and tubular flaring-lipped purplish or pink to magenta flowers. *Coldenia plicata* of the Colorado Desert hugs the ground like a mat, while *Coldenia greggii* found in the deserts of western Texas grows upright.

320 PURPLE MAT. *Nama demissum.* Low spreading annual of flats of clay or sand. Flowers vivid red-purple. Both California deserts, Arizona, and Baja California.

319

318

317

321 NOTCH-LEAF PHACELIA. *Phacelia crenulata.* Green, stout-stemmed annual with long-stamened, blue-purple flowers. The crushed herbage has a strong odor and stains the hands brown. Colorado and eastern Mohave deserts, east to New Mexico, south to Baja California.

322 FREMONT PHACELIA. *Phacelia fremontii.* A common species of sandy stretches and dry washes of the Mohave Desert and eastward into Nevada and Arizona. The handsome flowers, lavender-violet with yellow throat, have a somewhat skunky odor.

323 TANSY-LEAVED PHACELIA. *Phacelia tanacetifolia.* An annual often growing in the half-shade of creosote and cat's-claw bushes. The delicate foliage and blue flowers with well-exserted stamens make the plant most attractive. Colorado and Mohave deserts.

324 CLIMBING MILKWEED. *Funastrum heterophyllum.* Perennial, milky-juiced vine growing over the ground or up through, or over, bushes and trees; the flowers are purplish. Southern Nevada, western Arizona, and southeastern California, thence into Mexico.

325 SPANISH NEEDLE. *Palafoxia linearis.* Erect annual herb of sandy flats and dunes. The flowers are pinkish purple. The sap is exceedingly bitter. Southern Utah to western Arizona, southeastern California, and northwestern Mexico. A large coarse form (var. *gigantea*), up to 6 feet high and with larger leaves and heads, is found on the Algadones Dunes near Yuma, Arizona. A very handsome large-rayed species should be looked for on the Mexican sand dunes south of El Paso, Texas.

326 GIANT FOUR-O'-CLOCK. *Mirabilis froebelii.* A perennial herb of stony mesas, desert washes, and mountains of the Colorado and Mohave deserts. It often forms large "mats," "as big as a wagon wheel." Flowers bright purple. The large, night-blooming flowers are sweet scented.

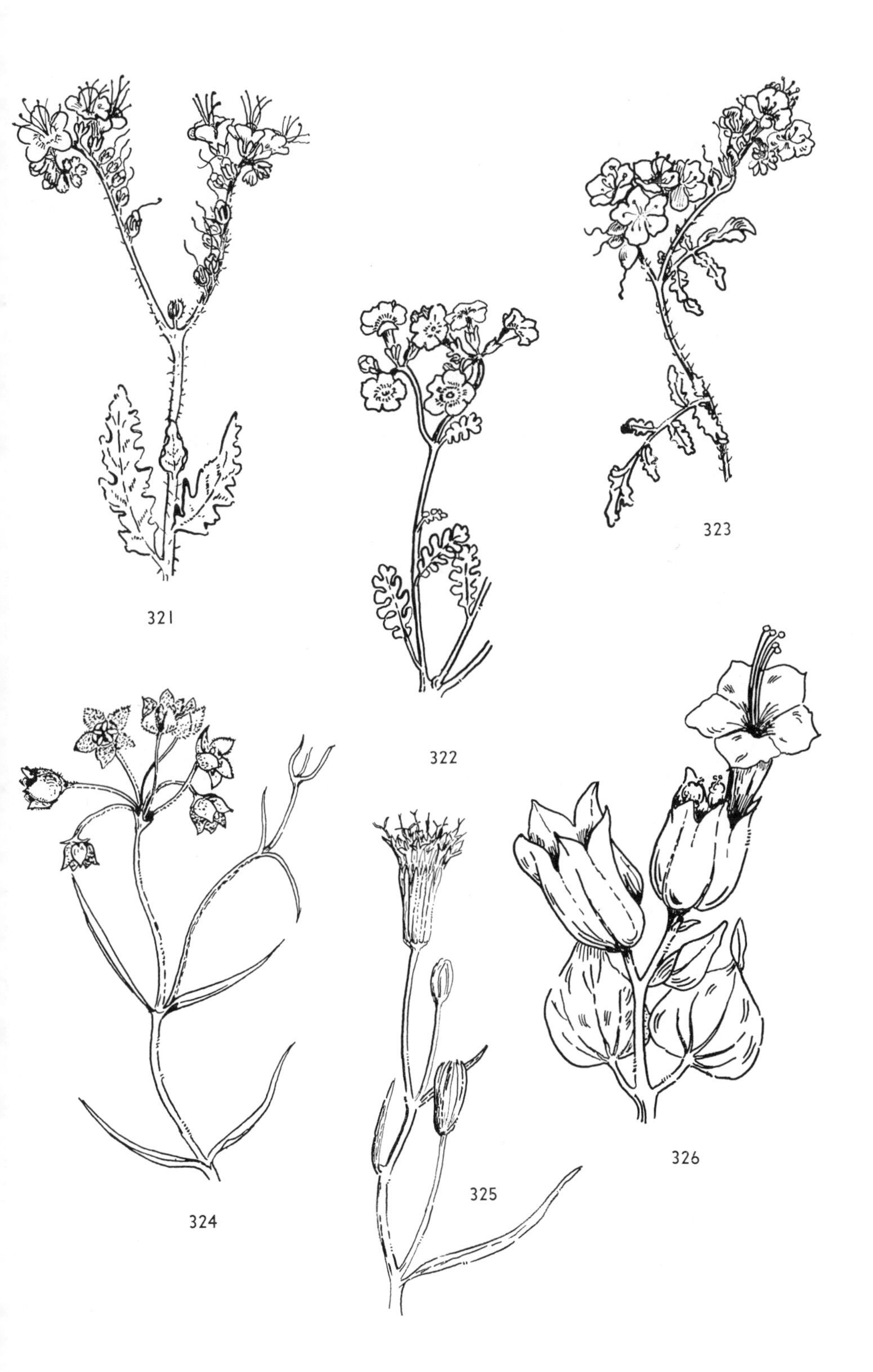
321
322
323
324
325
326

327 GREAT BASIN BLUE SAGE. *Salvia carnosa.* Handsome, erect, branched shrub, 1 to 2 feet high, often found with three-toothed sagebrush and juniper. The blue flowers contrast beautifully with the velvety purplish bracts and gray-green foliage. Washington to Arizona, California, and northern Baja California.

328 CHIA. *Salvia columbariae.* Upright annual with heads of flowers one above the other on the square-stemmed stalk. The small flowers are deep blue. The numerous small seeds are smooth. Indians of the California and Arizona deserts gathered the seeds in quantities for food. Southern California, southern Nevada, southwestern Utah, Arizona, and northern Baja California.

329 SNOWY BUSH MINT. *Poliomintha incana.* Silvery-leafed half-shrub, up to 3 feet tall, with numerous erect slender branches. Higher deserts. Sandy plains and desert mountain slopes of Colorado plateau, Tularosa Basin in New Mexico, and western Texas to Chihuahua in Mexico.

330 PAPER-BAG BUSH. *Salazaria mexicana.* Handsome rounded shrub, conspicuous because of its inflated, papery, often rose-tinted pods. Rocky hillsides and washes of Colorado and Mohave deserts, to Utah and Mexico. Upper flower-lip velvety blue, lower lip whitish.

331 MOHAVE GILIA. *Gilia densifolia mohavensis.* A pale-blue, perennial, several-stemmed gilia of rocky areas of the higher western and southern Mohave Desert. Superficially much like several other arid-land gilias such as *Gilia eremica* of sandy areas.

332 THISTLE SAGE. *Salvia carduacea.* One of the handsomest annuals of the Colorado and Mohave deserts, often occurring in dense stands. The herbage is white and woolly. Stalks 10 to 18 inches high. Flowers lavender with lacquer-red stamens.

333 DESERT LAVENDER. *Hyptis emoryi.* An early-blooming, ashy-leaved shrub of the lower rocky slopes and canyons of the Colorado Desert, southern Arizona, and adjacent Mexico. A fragrant shrub much visited by bees. A favorite nesting site for the verdin.

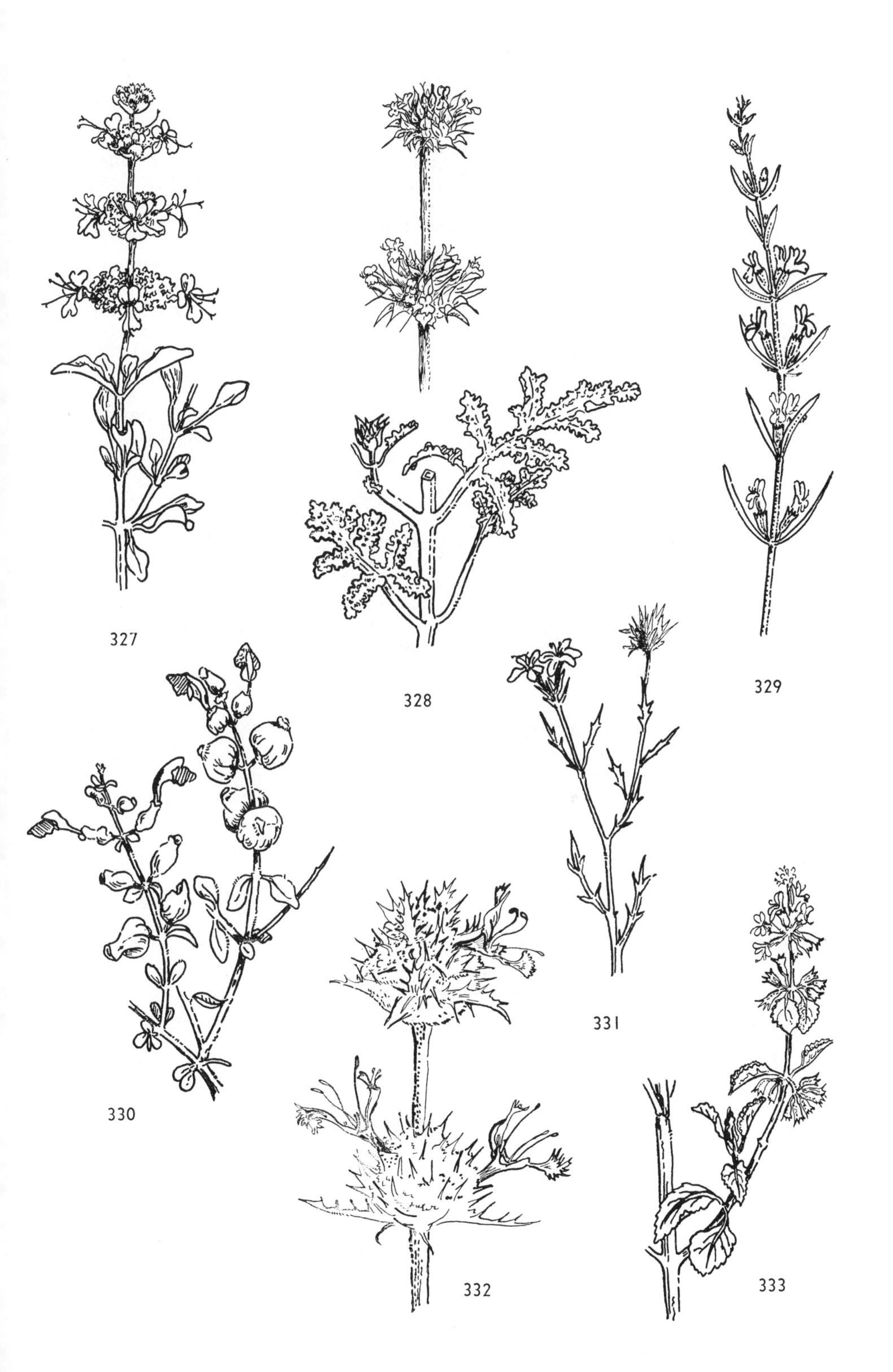

327
328
329
330
331
332
333

334 DESERT BLUEBELL. *Phacelia campanularia.* A handsome plant often occurring in groups. The deep blue flowers are most attractive. Valleys and mountains of the Colorado and Mohave desert, 1,000 to 4,000 feet.

335 RUELLIA. *Ruellia californica.* Shrub, 3 feet high or less, with handsome blue flowers and leaves covered with fine hairs. Grows generally along the edge of sandy washes. Sonora and Baja California. Belongs to the same family (*Acanthus*) as the BELOPERONE of the Colorado Desert.

336 CRUCIFIXION THORN. *Holacantha emoryi.* Small tree-like shrub, 6 to 12 feet high, with stout thorny branches and clusters of small reddish-purple to yellow flowers and small dry, nut-like fruits. Found as a rule on fine-textured soils of bottom lands along the lower Colorado and Gila River drainage south into northern Sonora. Sometimes confused with CANOTIA or KOEBERLINIA, which is also called crucifixion thorn.

337 HOPSAGE. *Grayia spinosa.* Low, stiff-branched, somewhat spiny shrub with slightly fleshy leaves and fruits subtended by conspicuous purple-reddish bractlets. A valuable browse plant of the Mohave and Great Basin deserts.

338 THURBER'S BEARD-TONGUE. *Penstemon thurberi.* A somewhat woody, several-to-many-stemmed perennial of open sandy or rocky slopes (2,000 to 4,000 feet). The flowers are blue-purple to pinkish. New Mexico to southern California, Baja California.

339 MEXICAN RABBIT-THORN or FRUTILLA. *Lycium brevipes.* Rigid spiny shrub, with lavender flowers and small green, several-seeded, tomato-like fruits. Colorado Desert south into Mexico. One of the many species of DESERT THORN, most of which have pinkish or lavender to purple, tubular, funnel-form flowers and semisucculent leaves.

340 FREMONT THORNBUSH. *Lycium andersonii.* Shrub, 5–6 feet high, with small succulent leaves and numerous pale-violet, trumpet-shaped flowers. The juicy small red fruits are much sought by quail. Alkaline soils of the Colorado Desert. It is one of many species of lyciums found in our southern deserts.

DESERT PLANTS (*blue*)

335

336

337

338

339

340

341 PARISH LARKSPUR. *Delphinium parishii.* A widely distributed plant of the Mohave and Colorado deserts. Flowers light blue.

342 DESERT HYACINTH. *Brodiaea capitata pauciflora.* Handsome long-leaved, tall-stemmed flower, widely scattered on our deserts from California to New Mexico. The bulbs, called grass-nuts, were eaten by early settlers and Indians.

343 DAVY GILIA. *Gilia davyi.* A fine showy species with tall stem and violet to pink flowers with yellow throat. Often in large aggregations in sandy soils of southern Nevada and the near-by California Mohave Desert.

344 WILD MORNING-GLORY. *Evolvulus arizonicus.* Erect, spreading, small leaved herb, with numerous fine stems and sky-blue, saucer-shaped flowers. The flowers vary much in size. Dry sandy plains and mesas. Arizona Upland Desert eastward to Chihuahuan Desert; southward into Sonora.

345 BLUE FLAX. *Linum lewisii.* Widely distributed perennial flax much resembling the annual cultivated flax of commerce. Occasional on deserts of southwestern United States and adjacent Mexico. Other kinds of flax with dainty yellow-to-orange flowers are also found on our deserts.

346 GUAYACAN. *Guaiacum coulteri.* Shrub or small tree, often with crooked branches; fragrant blue or violet flowers. The wood is hard, strong, and durable and often used as fuel. Sonora to Oaxaca.

347 DOTTED DALEA. *Dalea polyadenia.* Low-spreading, densely glandular shrub of plains and hillsides of the northern Mohave Desert to western Nevada.

348 SMOKE TREE. *Dalea spinosa.* A small, tree-like shrub of the larger sandy washes of the Sonoran Desert, especially in southern California, southern Arizona, and northern Baja California. It has few leaves, mostly thorns, to act as food-making organs. In early summer it is made handsome by multitudes of small deep-blue, pea-like flowers.

341

342

343

344

345

346

347

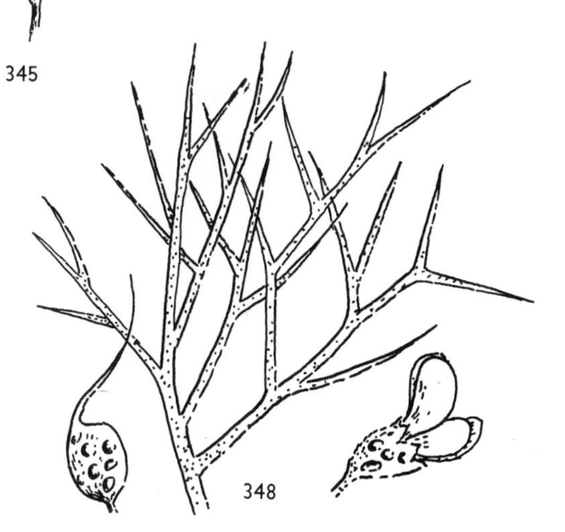
348

349 CANE CHOLLA. *Opuntia spinosior.* A rather tall, shrubby, very spiny cane-cactus with fleshy, solitary, bright lemon-yellow fruits persisting more than one season. The flowers are usually bright purple but occasionally they may be red, or even yellow or white. The numerous short spines are pinkish and give a characteristic color to the stems. The sheaths of the spines fall off after the first year. Sonora, New Mexico, and Arizona.

350 ENTRANA or JONCONOSTLE. *Opuntia imbricata.* A tall (6 to 12 feet), cylindrically jointed cholla with tree-like form, often forming extensive thickets. The large purple flowers are borne at the ends of the branches. The fruit is yellow and without spines. Central and northern Mexico.

351 FISHHOOK CACTUS. *Mammilaria tetrancistra.* A representative of a large group of cacti with teat-like tubercles covering the round or cylindrical stem. This one has both straight and hooked spines; there are others with only straight spines. Fishhook cactus has lavender flowers and bright-red fruits. Desert foothills from southwestern Utah to Arizona and southern California.

352 PITAHAYA AGRIA. *Machaerocereus gummosus.* A dark-green, long-stemmed cactus, common and often forming thickets on the desert plains throughout much of Baja California. The stems, at first erect, spread out horizontally, often bending and finally reaching the ground and taking root. The flowers are purple; the bright-scarlet fruits are agreeably tart and are much eaten by the natives.

353 INDIGO BUSH. *Dalea schottii.* An upright, 3-to-8-foot shrub of rocky washes, mesas, and hillsides. When crushed the foliage is very fragrant. In mid- and late-spring myriads of indigo-blue, pea-like flowers adorn the woody stems. Note the red oil pustules on fruits and leaves. Colorado Desert to Arizona and Baja California.

354 ROYAL LUPINE. *Lupinus odoratus.* Handsome, deep blue, fragrant-flowered, low-growing annual lupine of the higher deserts (3,000 to 4,500 feet) of western Arizona, southern Nevada, and southeastern California. Often in conspicuous dense aggregations.

355 DESERT IRONWOOD. *Olneya tesota.* A spreading tree of the hot southern deserts, attaining under favorable conditions a height of 27 feet. The bark is gray and the branches are armed with spines. The pea-like flowers are purple and white. The wood is brittle, very hard and heavy, and burns with a hot flame. The leaves may fall off during severe drought or after heavy frost. Found mostly in and about washes, from the southern Colorado Desert to southwestern Arizona, northwestern Sonora, and Baja California.

349

350

351

352

353

355

354

LIST OF PRONUNCIATIONS

agave ah-*gay*-vee
bolson bohl-*sohn*
calabazilla cah-lah-bah-*see*-yah
candelilla cahn-day-*lee*-yah
cardon cahr-*dohn*
chia *chee*-ah
cholla *choh*-yah
cirio *see*-ree-oh
copal *coh*-pahl
crucillo croo-*see*-yoh
entrana ayn-*trahn*-yah
flor de pena flohr day *pay*-nah
gallinita gah-yee-*nee*-tah
Gila *hee*-lah
guayacam gwah-yah-*cahm*
guayule gwah-*yoo*-lay
hojase hoh-*yah*-say
hote *hoh*-tay
huero de gato *whay*-roh day *gah*-toh
jito *hee*-toh
jojoba hoh-*hoh*-bah
lechuguilla lay-choo-*gee*-yah (hard "g")
malpais mahl-pah-*ees*
mesquite may-*skeet*
Mohave moh-*hah*-vay
ocotillo oh-coh-*tee*-yoh
palmilla pahl-*mee*-yah
palmito pahl-*mee*-toh
palo blanco *pah*-loh *blahn*-coh
palo verde *pah*-loh *vayr*-day
papache pah-*pah*-chay
piñon *peen*-yohn
piohito pee-oh-*ee*-toh
pitahaya pee-tah-*hah*-yah
sahuaro sah-*wah*-roh
senita say-*nee*-tah
sotol *soh*-tohl
tasajilla tah-sah-*jee*-yah
tornillo tohr-*nee*-yoh
torote tohr-*oh*-tay
Vizcaíno Vees-*cah*-ee-noh

BIBLIOGRAPHY

General References

Biology of Deserts. London: Institute of Biology, 1954.

BURT, W. H., and GROSSENHEIDER, R. P. *Field Guide to the Mammals.* Boston: Houghton Mifflin Co., 1952.

CAHALANE, VICTOR H. *Mammals of North America.* New York: Macmillan Co., 1947.

Desert Research. Research Council of Israel, Special Publications, No. 2. Jerusalem: Research Council of Israel, 1953.

DICE, LEE R. *The Biotic Provinces of North America.* Ann Arbor: University of Michigan Press, 1943.

The Future of Arid Lands. American Association for the Advancement of Science, Publication No. 43. Washington, D.C., 1956.

JAEGER, EDMUND C. *Desert Wild Flowers.* Rev. ed., Stanford, California: Stanford University Press, 1941.

———. *Our Desert Neighbors.* Stanford, California: Stanford University Press, 1950.

PETERSON, ROGER T. *A Field Guide to Western Birds.* Boston: Houghton Mifflin Co., 1941.

SHREVE, FORREST. *The Cactus and Its Home.* Baltimore: Williams & Wilkins Co., 1931.

STANDLEY, PAUL. *Trees and Shrubs of Mexico.* Contributions from the United States National Herbarium, No. 23, Parts 1–5. Washington, D.C.: Government Printing Office, 1920–24.

Chapter 3. The Chihuahuan Desert

GOLDMAN, EDWARD A., and MOORE, ROBERT T. "The Biotic Provinces of Mexico," *Journal of Mammalogy,* 26 (1946), 347–60.

LESUEUR, HARDE. *The Ecology of the Vegetation of Chihuahua, Mexico, North of Parallel Twenty-eight.* University of Texas Publications, Bulletin No. 4521. Austin: University of Texas, 1945.

MULLER, CORNELIUS H. "Vegetation and Climate of Coahuila, Mexico," *Madroño,* 9 (1947), 33–57.

SHREVE, FORREST. "Observations on the Vegetation of Chihuahua," *Madroño,* 5 (1939), 1–12.

Chapter 5. The Desert Plains and Foothills of Sonora

ALLEN, M. J. *Report on a Collection of Amphibians and Reptiles from Sonora, Mexico.* Occasional Papers, Museum of Zoology, University of Michigan, No. 259. Ann Arbor: University of Michigan, 1933.

BOLTON, HERBERT EUGENE (ed.). *Kino's Historical Memoir of Pimería Alta.* Berkeley: University of California Press, 1948.

BURT, WILLIAM HENRY. *Faunal Relationships and Geographical Distribution of Mammals in Sonora, Mexico.* Miscellaneous Publications, Museum of Zoology, University of Michigan, No. 39. Ann Arbor: University of Michigan, 1938.

HORNADAY, W. T. *Camp-fires on Desert and Lava.* New York: Charles Scribner's Sons, 1909.

JOHNSTON, IVAN. "Expedition to the Gulf of California, 1921," *Proceedings of the California Academy of Sciences,* Vol. XII, No. 30, pp. 951–1218. San Francisco, 1924.

SHREVE, FORREST, and WIGGINS, IRA L. *Vegetation and Flora of the Sonora Desert.* Carnegie Institution of Washington Publication No. 591, Vol. I. Washington, D.C., 1951.

Chapter 6. The Arizona Upland or Sahuaro Desert

BRANDT, HERBERT. *Arizona and Its Bird Life.* Cleveland: Bird Research Foundation, 1951.

BRYAN, KIRK. *The Papago Country, Arizona.* United States Geological Survey Water-Supply Paper No. 499. Washington, D.C.: Government Printing Office, 1925.

DI PESO, CHARLES. *The Upper Pima of San Cayetano del Tumacacori.* Dragoon, Arizona: The Amerind Foundation, Inc., 1956.

HOWES, PAUL GRISWOLD. *The Giant Cactus Forest and Its World.* Boston: Little, Brown and Co., 1954.

JOSEPH, ALICE, and SPICER, ROSAMOND B., and CHESKY, JANE. *The Desert People: A Study of the Papago Indians of Southern Arizona.* Chicago: University of Chicago Press, 1949.

Chapter 8. The Yuman Desert

BOLTON, HERBERT EUGENE (ed.). *Kino's Historical Memoir of Pimería Alta.* Berkeley: University of California Press, 1948.

Chapter 9. The Colorado Desert

CHASE, J. SMEATON. *California Desert Trails.* Boston: Houghton Mifflin Co., 1919.

GRINNELL, JOSEPH. *An Account of the Birds and Mammals of the Lower Colorado Valley.* University of California Publications in Zoology, XV. Berkeley: University of California Press, 1915.

MENDENHALL, W. C. *Groundwaters of the Indio Region, California.* United States Geological Survey Water-Supply Paper No. 225. Washington, D.C.: Government Printing Office, 1909.

REMPEL, PETER J. "The Crescentic Dunes of the Salton Sea and Their Relation to Vegetation," *Ecology,* 17 (1936), 347–58.

SHREVE, FORREST. "The Transition from Desert to Chaparral in Baja California," *Madroño,* 3 (1936), 257–64.

SYKES, GODFREY. "The Delta and Estuary of the Colorado River," *Geographical Review,* 16 (1926), 232-55.

Chapter 10. The Vizcaíno-Magdalena Desert

BAEGERT, JOHANN JAKOB. *Observations in Lower California.* Berkeley: University of California Press, 1951.

BEAL, CARL H. *Reconnaissance of the Geology and Oil Possibilities of Baja California, Mexico.* Geological Society of America Memoir No. 31. Baltimore: Geological Society of America, 1948.

GERHARD, PETER, and GULICK, HOWARD E. *Lower California Guidebook.* Glendale, California: Arthur H. Clark Co., 1956.

GOLDMAN, E. A. *Plant Records of an Expedition to Lower California.* Contributions from the United States National Herbarium, No. 16. Washington, D.C.: Government Printing Office, 1916.

GRINNELL, JOSEPH. *A Distributional Summation of the Ornithology of Lower California.* University of California Publications in Zoology, XII. Berkeley: University of California Press, 1921.

JOHNSTON, IVAN M. *Expedition of the California Academy of Sciences to the Gulf of California in 1921: Botany (The Vascular Plants).* San Francisco: California Academy of Sciences, 1924.

MEIGS, P. *The Dominican Mission Frontier of Lower California.* Berkeley: University of California Press, 1935.

NELSON, E. W. *Lower California and Its Natural Resources.* Memoirs of the United States National Academy of Sciences, No. 16, Washington, D.C.: Government Printing Office, 1921.

NORDOFF, CHARLES. *Peninsular California.* New York: Harper & Brothers, 1888.

NORTH, ARTHUR WALBRIDGE. *Camp and Camino in Lower California.* New York: Baker and Taylor Co., 1916.

———. *The Mother of California.* San Francisco and New York: Paul Elder and Co., 1908.

Chapter 12. The Mohave Desert

DARTON, N. H., and OTHERS. *Guidebook of the Western United States: Part C. The Santa Fe Route.* United States Geological Survey Bulletin No. 613. Washington, D.C.: Government Printing Office, 1916.

JAEGER, EDMUND C. *The California Deserts.* Rev. ed., Stanford, California: Stanford University Press, 1955.

KIRK, RUTH. *Exploring Death Valley.* Stanford, California: Stanford University Press, 1956.

MENDENHALL, WALTER C. *Some Desert Watering Places in Southeastern California and Southwestern Nevada.* Washington, D.C.: Government Printing Office, 1909.

MILLER, WILLIAM J. "Geology of Parts of the Barstow Quadrangle, San Bernardino County, California," *California Journal of Mines and Geology,* 40 (1944).

THOMPSON, DAVID G. *The Mohave Desert Region.* United States Geological Survey Water-Supply Paper No. 578. Washington, D.C.: Government Printing Office, 1929.

Chapter 13. The Great Basin Desert

BAILEY, FLORENCE MERRIAM. *Handbook of Birds of the Western United States.* Rev. ed., Boston: Houghton Mifflin Co., 1921.

BLACKWELDER, E., HUBBS, CARL, MILLER, ROBERT, and ANTEVS, ERNEST. *The Great Basin.* Bulletin of the University of Utah, Biological Series, Vol. X, No. 7, Salt Lake City: University of Utah, 1948.

BLAKE, WILLIAM P. *Geological Report: Explorations and Surveys to Ascertain the Most Practical and Economical Route for a Railroad from the Mississippi River to the Pacific Ocean.* Washington, D.C., 1857.

BOUTWELL, JOHN M. *The Salt Lake Region.* Sixteenth International Geological Congress, Guidebook No. 17, Excursion C-1, pp. 1–149. Washington, D.C.: Government Printing Office, 1932.

CHAPMAN, H. R. "The Deserts of Nevada and Death Valley," *National Geographic Magazine*, 17 (1906), 483–97.

DAVIS, WILLIAM MORRIS. "The Mountain Ranges of the Great Basin," *Bulletin of the Museum of Comparative Zoology*, 42 (1903), 129–77.

HALL, E. RAYMOND. *Mammals of Nevada.* Berkeley: University of California Press, 1946.

HOUGHTON, ELIZABETH DONNER. *The Expedition of the Donner Party and Its Tragic Fate.* Chicago: A. C. McClurg and Co., 1911.

OBERHOLSER, HARRY C. "Glimpses of Bird Life in the Great Basin," *Annual Report of the Smithsonian Institution*, 1919, pp. 355–66. Washington, D.C.: Government Printing Office, 1921.

STANSBURY, HOWARD. *Exploration and Survey of the Valley of the Great Salt Lake of Utah.* Philadelphia: Lippincott, Grambo & Co., 1852.

Chapter 14. The Painted Desert

COLTON, HAROLD S., and BAXTER, F. C. *Days in the Painted Desert and the San Francisco Mountains, Arizona—A Guide.* Museum of Northern Arizona Bulletin No. 2. Flagstaff, Arizona, 1932.

DARTON, N. H., and OTHERS. *Guide Book of the Western United States: Part C. The Santa Fe Route.* United States Geological Survey Bulletin No. 613. Washington, D.C.: Government Printing Office, 1916.

LUOMALA, KATHARINE. *Navaho Life of Yesterday and Today.* Berkeley, California: United States Department of the Interior, National Park Service, 1933.

SIMPSON, RUTH D. *The Hopi Indians.* Southwest Museum Leaflets, No. 25. Los Angeles, 1953.

SWANTON, JOHN R. *The Indian Tribes of North America.* Bureau of American Ethnology Bulletin No. 145. Washington, D.C.: Government Printing Office, 1950.

WENDORF, FRED. *Archeological Studies in the Petrified Forest National Monument, Arizona.* Museum of Northern Arizona Bulletin No. 27. Flagstaff, Arizona, 1953.

INDEX